AF353117

Advances in Supercapacitor Technology and Applications

Advances in Supercapacitor Technology and Applications

Special Issue Editors

Alon Kuperman
Alessandro Lampasi

MDPI • Basel • Beijing • Wuhan • Barcelona • Belgrade • Manchester • Tokyo • Cluj • Tianjin

Special Issue Editors

Alon Kuperman
Ben-Gurion University of the Negev
Israel

Alessandro Lampasi
National Agency for New Technologies, ENEA
Italy

Editorial Office
MDPI
St. Alban-Anlage 66
4052 Basel, Switzerland

This is a reprint of articles from the Special Issue published online in the open access journal *Energies* (ISSN 1996-1073) (available at: https://www.mdpi.com/journal/energies/special_issues/supercapacitor).

For citation purposes, cite each article independently as indicated on the article page online and as indicated below:

LastName, A.A.; LastName, B.B.; LastName, C.C. Article Title. *Journal Name* **Year**, *Article Number*, Page Range.

ISBN 978-3-03936-388-9 (Hbk)
ISBN 978-3-03936-389-6 (PDF)

Contents

About the Special Issue Editors . . . . **vii**

S. Srinivasa Rao, Ikkurthi Kanaka Durga, Bandari Naresh, Bak Jin-Soo, T.N.V. Krishna, Cho In-Ho, Jin-Woo Ahn and Hee-Je Kim
One-Pot Hydrothermal Synthesis of Novel Cu-MnS with PVP Cabbage-Like Nanostructures for High-Performance Supercapacitors
Reprinted from: *Energies* **2018**, *11*, 1590, doi:10.3390/en11061590 . . . . **1**

Yedluri Anil Kumar and Hee-Je Kim
Effect of Time on a Hierarchical Corn Skeleton-Like Composite of CoO@ZnO as Capacitive Electrode Material for High Specific Performance Supercapacitors
Reprinted from: *Energies* **2018**, *11*, 3285, doi:10.3390/en11123285 . . . . **15**

Anil Kumar Yedluri, Tarugu Anitha and Hee-Je Kim
Fabrication of Hierarchical NiMoO$_4$/NiMoO$_4$ Nanoflowers on Highly Conductive Flexible Nickel Foam Substrate as a Capacitive Electrode Material for Supercapacitors with Enhanced Electrochemical Performance
Reprinted from: *Energies* **2019**, *12*, 1143, doi:10.3390/en12061143 . . . . **31**

Anil Kumar Yedluri, Eswar Reddy Araveeti and Hee-Je Kim
Facilely Synthesized NiCo$_2$O$_4$/NiCo$_2$O$_4$ Nanofile Arrays Supported on Nickel Foam by a Hydrothermal Method and Their Excellent Performance for High-Rate Supercapacitance
Reprinted from: *Energies* **2019**, *12*, 1308, doi:10.3390/en12071308 . . . . **43**

Lujun Wang, Jiong Guo, Chen Xu, Tiezhou Wu and Huipin Lin
Hybrid Model Predictive Control Strategy of Supercapacitor Energy Storage System Based on Double Active Bridge
Reprinted from: *Energies* **2019**, *12*, 2134, doi:10.3390/en12112134 . . . . **55**

Fernando Ortenzi, Manlio Pasquali, Pier Paolo Prosini, Alessandro Lidozzi and Marco Di Benedetto
Design and Validation of Ultra-Fast Charging Infrastructures Based on Supercapacitors for Urban Public Transportation Applications
Reprinted from: *Energies* **2019**, *12*, 2348, doi:10.3390/en12122348 . . . . **75**

Jianhua Guo, Weilun Liu, Liang Chu and Jingyuan Zhao
Fractional-Order Modeling and Parameter Identification for Ultracapacitors with a New Hybrid SOA Method
Reprinted from: *Energies* **2019**, *12*, 4251, doi:10.3390/en12224251 . . . . **89**

Evgeny P. Kitsyuk, Renat T. Sibatov and Vyacheslav V. Svetukhin
Memory Effect and Fractional Differential Dynamics in Planar Microsupercapacitors Based on Multiwalled Carbon Nanotube Arrays
Reprinted from: *Energies* **2020**, *13*, 213, doi:10.3390/en13010213 . . . . **103**

Gustavo Navarro, Jorge Nájera, Jorge Torres, Marcos Blanco, Miguel Santos and Marcos Lafoz
Development and Experimental Validation of a Supercapacitor Frequency Domain Model for Industrial Energy Applications Considering Dynamic Behaviour at High Frequencies
Reprinted from: *Energies* **2020**, *13*, 1156, doi:10.3390/en13051156 . . . . **117**

Mostafa Kermani, Giuseppe Parise, Ben Chavdarian and Luigi Martirano
Ultracapacitors for Port Crane Applications: Sizing and Techno-Economic Analysis
Reprinted from: *Energies* **2020**, *13*, 2091, doi:10.3390/en13082091 . . . . **135**

About the Special Issue Editors

Alon Kuperman received the Ph.D. degree in electrical and computer engineering from the Ben-Gurion University of the Negev, Beersheba, Israel, in 2006. He was a Marie Curie Training Site Member with Imperial College London, London, U.K, in 2003–2006. He was an Honorary Research Fellow with the University of Liverpool, Liverpool, U.K., in 2008–2009. He is currently with the School of Electrical and Computer Engineering, Ben-Gurion University, Israel, where he is heading the Power and Energy Systems track and directing the Applied Energy Laboratory. His research interests include all aspects of energy conversion and applied control.

Alessandro Lampasi received the M.Sc. degree in electronic engineering and the Ph.D. degree in electrical engineering from the University of Rome Sapienza, Rome, Italy, in 2002 and 2006, respectively. He is currently with the Italian National Agency for New Technologies, Energy and Sustainable Economic Development (ENEA), Frascati, Italy, where he is in charge for several international projects concerning nuclear fusion and power systems. His current research interests include modeling and measurement techniques in the fields of nuclear fusion, power electronics, energy storage, and applied electromagnetics.

Article

One-Pot Hydrothermal Synthesis of Novel Cu-MnS with PVP Cabbage-Like Nanostructures for High-Performance Supercapacitors

S. Srinivasa Rao [1], Ikkurthi Kanaka Durga [2], Bandari Naresh [2], Bak Jin-Soo [2], T.N.V. Krishna [2], Cho In-Ho [2], Jin-Woo Ahn [1] and Hee-Je Kim [2,*]

[1] Department of Mechatronics Engineering, Kyungsung University, 309 Suyeong-ro, Nam-gu, Busan 48434, Korea; srinu.krs@gmail.com (S.S.R.); jwahn@ks.ac.kr (J.-W.A.)

[2] School of Electrical Engineering, Pusan National University, Busandaehak-ro 63beon-gil, Geumjeong-gu, Busan 46241, Korea; kanakadurga.ikkurthi@gmail.com (I.K.D.); naresh1.bandari@gmail.com (B.N.); imjsn17@pusan.ac.kr (B.-J.S.); vamsik.tirumalasetty@gmail.com (T.N.V.K.); whdlsgh92@icloud.com (C.I.-H.)

* Correspondence: heeje@pusan.ac.kr; Tel.: +82-51-510-2364; Fax: +82-51-513-0212

Received: 28 April 2018; Accepted: 13 June 2018; Published: 17 June 2018

Abstract: This paper reports the facile synthesis of a novel architecture of Cu-MnS with PVP, where the high theoretical capacitance of MnS, low-cost, and high electrical conductivity of Cu, as well as appreciable surface area with high thermal and mechanical conductivity of PVP, as a single entity to fabricate a high-performance electrode for supercapacitor. Benefiting from their unique structures, the Cu-MnS with 2PVP electrode materials show a high specific capacitance of 833.58 F g^{-1} at 1 A g^{-1}, reversibility for the charge/discharge process, which are much higher than that of the MnS-7 h, Cu-MnS, and Cu-MnS with 1 and 3PVP. The presence of an appropriate amount of PVP in Cu-MnS is favorable for improving the electrochemical performance of the electrode and the existence of Cu was inclined to enhance the electrical conductivity. The Cu-MnS with 2PVP electrode is a good reference for researchers to design and fabricate new electrode materials with enhanced capacitive performance.

Keywords: supercapacitors; cabbage plant like nanostructured; Cu-MnS with PVP; electrochemical studies; stability

---

1. Introduction

Research related to energy production, storage, and distribution is becoming increasingly important due to the large proliferation of portable electronics, auxiliary power sources, and hybrid electric vehicles in recent years [1,2]. The two most frequently used energy storage devices are lithium ion batteries and electrolytic capacitors. Lithium ion batteries have a very low power density with low cycle life and suffering from ageing and it can explode due to overcharge [3,4]. In contrast, electrolytic capacitors also can explode due to the spark ignition of free oxygen or high gas pressures and it shows a low energy density [5]. Therefore, both commonly used energy storage devices are not secure and suitable when high energy and power densities are required. From an energy storage perspective, supercapacitors with high power densities of 10 kW kg^{-1} with excellent energy densities of approximately 150–200 W h kg^{-1} are the prominent options [6]. Depending on their energy storage mechanism, supercapacitors are categorized mainly into three groups: electric double layer capacitors (EDLC), pseudocapacitors, and hybrid electrochemical capacitors. EDLC consisting of carbon based materials such as carbon nanotubes, carbon hollow nanospheres, activated carbon, graphene, nitrogen-doped carbon nanofibers, and storage charge via the electrostatic interactions of ions at the electrode/electrolyte interface, and thus exhibit relatively low specific capacitance (SC); however, hybrid electrochemical capacitors usually consist of one battery-type Faradaic electrode with another capacitive electrode in the entire device [7,8]. Pseudocapacitors usually consisting of metal oxide/hydroxide or conducting polymer electrodes display higher energy density and higher specific capacitance, and redox reactions take place at the surface of the electroactive materials. Therefore, pseudocapacitors have attracted considerable research interest in recent years and exhibited a higher SC (300–1000 F g^{-1}) than that of EDLC [9]. Pseudocapacitors have many other benefits, such as excellent performance and high power density, compared to traditional lithium ion batteries. Therefore, pseudocapacitors are promising devices for low cost, easy to maintain, fast charging, small size, extensive range of setup temperatures, high-performance energy storage, eco-friendly, and safe applications [10,11].

On the other hand, many studies of traditional pseudocapacitor materials have been limited by the lack of high performance with a facile preparation method or reasonable cost. In addition, the electrochemical stability or cycling performance of polymer composite electrodes is challenging with a >10% reduction of SC after 1000 charging/discharging cycles. Moreover, the high-performance energy storage electronic devices currently used in industry are susceptible to accidental fire damage due to electrolyte leakage, overheating, and current overflow [12,13]. These issues have prevented their wider use by traditional companies manufacturing modern electronic devices. These problems can resolved and these devices can be made secure by synthesizing hybrid ternary nanocomposites with pseudocapacitor materials.

Transition metal nitrides (TMNs) have been investigated widely for supercapacitor applications. On the other hand, the synthesis of TMNs is a very difficult and time consuming process due to the formation of N≡N bonds [14]. Moreover, it shows instability in humid atmospheres and air, which can drastically alter the physicochemical properties and turn the energy storage performance of the material. Among the metal sulfides candidates, manganese sulfide (MnS) exhibits intriguing characteristics, such as low-cost, exceptional electrochemical performance, environmental friendliness, and natural abundance [15,16]. This is the most promising electrode material for the next generation supercapacitors because of its high theoretical SC of 1370 F g^{-1} in aqueous electrolytes. Compared to ternary transition metal oxides, metal sulfides often exhibit higher conductivity, ionic diffusivity compared to metal oxides due to the substitution of oxygen with sulfur atoms, narrow band gap, large anionic polarizability, and larger sizes of the S^{2-} ion [15,17]. On the other hand, regardless of its high theoretical SC value, the experimental SC of MnS supercapacitors is much lower than that of the anticipated value owing to its meager electronic conductivity (10^{-5}–10^{-6} S cm^{-1}), low surface area of 10–80 m^2 g^{-1}, and poor cyclic stability. The SC value and electrical conductivity of the MnS electrode material can be enhanced further by synthesizing a binary composite with copper (Cu) and polyvinyl pyrrolidone (PVP). Composite materials possess synergistic improvements in properties, such as cycling stability, charge transfer resistance, and chemical activity to poisoning for SCs, which are greatly superior to a simple combination of individual components [18]. Among the many metallic materials, Cu has attracted significant interest because of its essential roles in metal-catalyzed reactions and its environmental credentials. In addition, Cu is of particular interest because of its high electrical conductivity and it is used widely in catalysts, resins, thermal conduction, and electronics [19].

Mochao Cai et al. reported the PVP-assisted synthesis of a Fe$_3$O$_4$/graphene composite for lithium storage applications [20]. The presence of PVP molecules stabilized the graphene oxide suspension and enhanced the reversible capacity, cycling performance, and rate capability significantly. Furthermore, PVP nanoparticles are the least toxic, have a high surface area, and show fast electron communication features; hence, they have been used extensively as nanosensors for a range of biomolecules and ions. Xumin Zhang et al. used PVP to modify graphene oxide for enhancing the thermal and mechanical conductivity and observed appreciable results [21,22].

This paper reports the facile synthesis of a novel architecture of Cu-MnS with PVP, where the complementary features of high theoretical capacitance of MnS, low-cost, and high electrical conductivity of Cu as well as the appreciable surface area with high thermal and mechanical conductivity of PVP, on a single entity on nickel foam to fabricate a high-performance electrode for supercapacitor applications. The active phase of Cu and PVP in MnS promotes ion diffusion on the interfaces, and the void space between the particles can endure the volume change in long-term cycling. The electrochemical tests revealed the as-prepared Cu-MnS with the 2PVP electrode to have an extraordinary SC of 833.58 F g^{-1} at a current density of 1 A g^{-1} and very high cycling stability of 96.95% over 1000 cycles in a 2 M KOH solution. Therefore, it can provide a new pathway towards the commercialization of Cu-MnS with PVP-based supercapacitors.

2. Experimental Methods

2.1. Materials

Manganese sulfate monohydrate [$MnSO_4 \cdot H_2O$], sodium sulfate [Na_2SO_4], ammonium persulfate [$(NH_4)_2S_2O_8$], copper sulfate pentahydrate [$CuSO_4 \cdot 5H_2O$], polyvinyl pyrrolidone, polyvinylidene fluoride, potassium hydroxide KOH, 1-methyl-2-pyrrolidinone were purchased from Sigma Aldrich. Carbon black acetylene was purchased from Alfa Aesar. All chemicals used in the experiment were of analytical grade and used as received.

2.2. Synthesis of Cu-MnS with PVP Nanoparticles

In a typical synthesis of MnS, 0.1 M $MnSO_4 \cdot H_2O$, 0.1 M Na_2SO_4, and 0.1 M of $(NH_4)_2S_2O_8$ were dissolved in 30 mL of deionized water with vigorous stirring. The solution was stirred for 30 min under ambient conditions to form a clear white color solution and then transferred to a Teflon lined stainless steel autoclave heated to 150 °C for 5 h. After cooling naturally to room temperature, the as-obtained MnS material was collected, filtered, and washed several times with deionized water and ethanol to remove the excess reactants and loosely bound ions. The black precipitates were collected and dried in an oven at 80 °C for 10 h to enhance the specific surface area. Cu-doped MnS and Cu-doped MnS with PVP were prepared using the same procedure, where 10 wt. % of $CuSO_4 \cdot 5H_2O$ and/or polyvinyl pyrrolidone (0.1, 0.2, and 0.3 g) were dispersed into the MnS mixture and heated to 150 °C for 5 h.

2.3. Preparation of Working Electrode on Nickel Foam Substrate

The nickel foam substrate with rectangular dimensions of 1×1 cm^2 was used as a substrate because of its good electrical conductivity and desirable 3D porous structure. The nickel foam reduces the diffusion resistance of the redox electrolyte and improves the facility of ion transportation [23,24]. Prior to synthesis, the nickel foam substrates was cleaned with acetone and a 3 M HCl solution for 10 min to remove the surface oxide layer. Finally, the nickel foam was washed with ethanol and deionized water by sonication and dried with a hair dryer. The working electrode was prepared by mixing 75 wt.% of active material, 15 wt.% of carbon black acetylene, and 5 wt.% of polyvinylidene fluoride in 2 mL of 1-methyl-2-pyrrolidinane with stirring overnight. The resulting slurry was loaded on the nickel foam and dried overnight in an oven at 60 °C. The loading densities of the active materials were acquired by measuring the electrode with a microbalance with an accuracy of 0.01 mg and the weight of the active materials was approximately 5 mg cm^{-2} for all electrodes. The loading was measured by calculating the difference between the measured mass of the active materials coated and uncoated on a nickel foam substrate.

2.4. Characterizations

The resulting MnS, Cu-MnS, and Cu-MnS with PVP electrodes were characterized by field emission scanning electron microscopy (FE-SEM; SU-70, Hitachi, Tokyo, Japan). Their compositions and crystalline structures were characterized by energy-dispersive X-ray spectroscopy (EDX) attached to the FE-SE microscope and powder X-ray diffraction (XRD, Rigaku D/MAX 2500PC, Rigaku Corporation, Tokyo, Japan), which were collected from 20 to 80 °C in 2θ with a Cu target and a monochromator at 40 kV and 250 mA. X-ray photoelectron spectroscopy (XPS, KBSI, Busan, Korea) was performed to examine the elemental compositions using Al Kα X-rays at 240 W.

2.5. Electrochemical Characterization

The electrochemical tests were carried out in a classical three-electrode setup; hierarchical Cu-MnS with PVP/nickel foam architectures were used directly as the working electrode, platinum wire, and Ag/AgCl as a counter electrode and reference electrode. Cyclic voltammetry (CV, SP-150 Biological Science Instrument, Seyssinet-Pariset, France) was performed at room temperature over the voltage range of −0.5 to 0.5 V at different scan rates using a three and two-electrode cell. The galvanostatic charge–discharge (GCD) cycles were measured within a voltage range of 0.4 to −0.5 V at various current densities to evaluate the SC. To evaluate the SC retention, three thousand cycles were performed at a high current density and the GCD curves were derived. The specific capacitance, area capacitance, and the energy and power densities values was calculated for the single electrodes and their respective equations are given in the supporting information in Sections 1–3. Electrochemical impedance spectroscopy (EIS) was observed using a biologic potentiostat/galvanostat/EIS analyzer (Biological Science Instrument, Seyssinet-Pariset, France) by applying frequencies from 100 KHz to 0.1 Hz and an AC voltage of 5 mV at the open circuit potential.

3. Results and Discussion

Figure 1 presents the fabrication process of MnS, Cu-MnS, and Cu-MnS with PVP samples based on the experimental data. Published reports suggest that nanotubes, nanowires, and nanorods are the most common 1D nanostructures and they have attracted considerable attention due to facile strain relaxation during ion intercalation, efficient 1D electron transport and effective electrode-electrolyte contact because of their large surface-to-volume ratio [25,26]. Cu-MnS with PVP was prepared by mixing MnS, Cu, and PVP solution at specific ratios, as described in the experimental section to achieve efficient and stable materials. The cabbage-like nanostructured sample was prepared by a hydrothermal method at 150 °C for 5 h (Figure 1b). The resulting Cu-MnS with 2PVP wet powder (Figure 1c) was cleaned and filtered with DI water and ethanol and annealed further at 80 °C for 10 h. The powder was then coated on nickel foam and used further for electrochemical reactions.

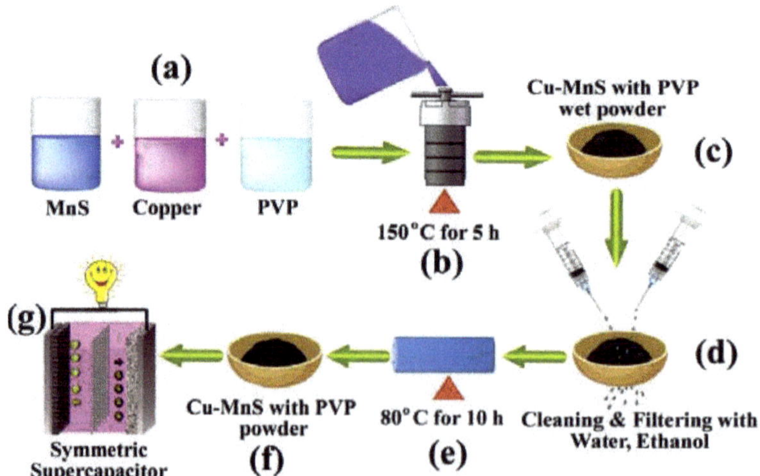

Figure 1. Illustration of the synthesis route for Cu-MnS with PVP nanocomposite.

The crystallographic structure of the product obtained by the hydrothermal method was determined by X-ray power diffraction (XRD). Figure 2a presents the XRD patterns of the MnS-7 h, Cu-MnS, and Cu-MnS with 2PVP nanocomposite. Figure 2a (MnS-7 h) showed broad XRD peaks at

28.53, 34.30, 40.72, 42.61, 56.43, 58.96, 64.79, 67.01, and 72.30° 2θ. No other characteristic peaks from impurities—such as $Mn(OH)_2$, MnO, or MnO_2—were detected, which reveals the high-phase purity of the as-prepared MnS. On the other hand, the intensity of the pure MnS peaks decreased gradually with the deposition of Cu and 2PVP in MnS and no XRD peaks were observed for Cu-MnS with 2PVP, indicating that the Cu and PVP were dispersed well in MnS. The observed values are accordance with the reported data (Joint Committee for Powder Diffraction studies file no. 40-1288 and 40-1289).

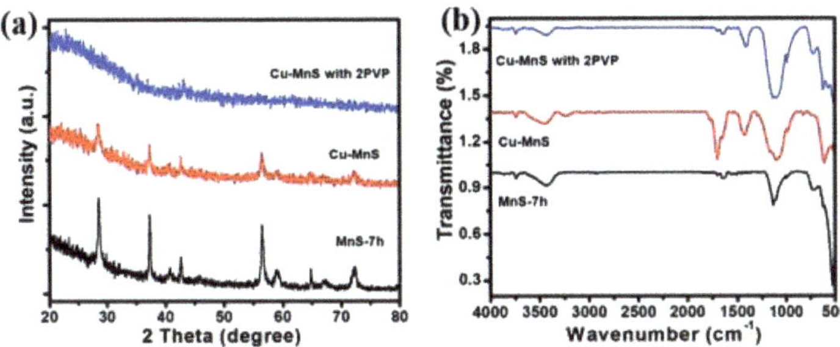

Figure 2. XRD (**a**) and FTIR (**b**) patterns of the MnS, Cu-MnS, and Cu-MnS with 2PVP nanocomposite.

The vibrational properties were also studied by Fourier transform infrared (FTIR) spectroscopy, as shown in Figure 2b. The peaks at approximately 646 and 823 cm^{-1} were attributed to the Mn-S stretching vibration and the existence of a response interaction between the vibrational modes of sulfide ions in the powder. The additional band observed at approximately 1253 cm^{-1} was assigned to the sulfide source in MnS, indicating the coordination between Mn atoms in MnS. The bands at 3000–3600 cm^{-1} were assigned to the –OH group, indicating the existence of H_2O absorbed in the surface of nanocrystals. The broad absorption peak in the range, 1500–1650 cm^{-1}, were assigned to the C=O stretching modes due to the absorption of carbon dioxide [27,28]. The FTIR spectrum of Cu-MnS and Cu-MnS with 2PVP samples displayed comparable peaks to the spectrum of pure MnS-7 h. The peak at 1253 cm^{-1} was split into multiple peaks, indicating that the doped Cu and PVP affected the structure of MnS.

XPS was conducted to evaluate the chemical states and metal oxidation states of the MnS, Cu-MnS and Cu-MnS with 2PVP materials. Figure 3 shows the corresponding results. All the analysis (XRD, XPS and FTIR spectroscopy) were conducted using the powder to exclude the effects of nickel foam substrate. As shown in Figure 3a, the wide-scan XPS survey spectra of the three different materials confirmed the presence of S2p, C1s, N1s, O1s, Mn2p, and Na1s. The Mn2p spectrum of Cu-MnS with 2PVP nanocomposite could be fitted to two binding energies of 642.00 and 653.85 eV, corresponding to the $Mn2p_{3/2}$ and $Mn2p_{1/2}$, respectively. The Mn2p for Cu-MnS with the 2PVP nanocomposite showed two strong intensity peaks, which were higher than those of the bare MnS-7 h and Cu-MnS, suggesting the strong formation of Mn in the composite [29]. The presence of Cu and PVP in MnS can increase the electrochemical activity of MnS in the electrochemical process. According to the binding energy separation of Mn2p, a ΔE of 11.85 eV was observed for the resulting materials indicating that the dominant valence of Mn is Mn^{4+}. Figure 3c presents the S2p spectra of the resulting electrodes; the major peak at 168.1 eV was assigned to the $S2p_{3/2}$ characteristic peak of sulfur in MnS. Copper and PVP were not detected in the Cu-MnS and Cu-MnS with 2PVP due to the low crystallinity and too thin layers of Cu and PVP. The new peak at approximately 163.00 eV in Figure 3c for Cu-MnS with 2PVP can be associated with the partial oxidation of unsaturated S atoms in air. Moreover, from XPS analysis of MnS-7 h, Cu-MnS, and Cu-MnS with 2PVP composites, the presence of Mn (17.97, 9.49, and 2.05%), S (4.22, 9.69, and 6.24% for the MnS, Cu-MnS, and Cu-MnS with 2PVP), O, and C was quite evident without impurities. The XPS curves showed that Mn and S were distributed uniformly and there was a high C content, as shown in Figure 3d.

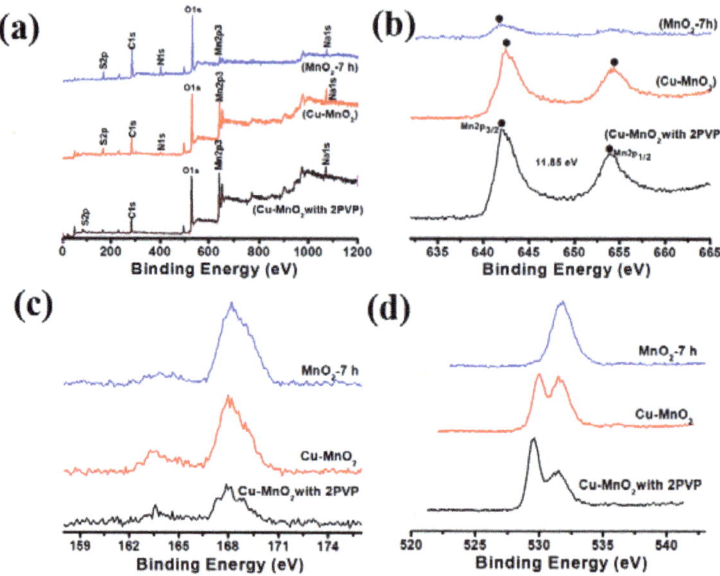

Figure 3. The XPS survey spectrum of (a) MnS-7 h, Cu-MnS, Cu-MnS with 2PVP and high resolution spectrums of (b) Mn2p, (c) S2p and (d) O1s of MnS-7 h, Cu-MnS and Cu-MnS with 2PVP composite.

In this study, nickel foam was used as a substrate that exhibits good electrical conductivity and a 3D hierarchical structure with open macropores that allow prompt ion and electron transfer. Figure 4a,c presents FE-SEM images of bare MnS and Cu-MnS grown on nickel foam, and Figure 4b,d show their corresponding magnified images. The surface of MnS consists of distinct wrinkled nanotubes that can be seen from their end tips (Figure 4b), which might have emerged during the hydrothermal cooling process due to a divergence in the coefficient of thermal expansion [30]. These wrinkled nanotubes can serve as nucleation sites for the growth of Cu-MnS with PVP nanostructures as well as allow the uniform distribution of Cu and PVP. After adding 10 wt.% of Cu into MnS, the surface morphology did not change substantially but the length (580–600 nm) and diameter (100 nm) of the nanotubes decreased, which revealed the agglomeration of wrinkled nanotubes.

The effect of PVP in Cu-MnS with three different concentrations on the surface morphologies of the samples was then evaluated, as shown in Figure 5. The Cu-MnS with 1PVP nanocomposite exhibited a combination of nanomushroom and nanoflower-like structured morphologies. The low magnification SEM images covering more than 10 µm in length, displayed many nanoflower like structures, which were not uniform in morphology and formed many micro/nanogaps. High resolution images (Figure 5b) show that the flower composed of number of interlaced flakes. Figure 5a,b shows that the cross-linked flakes agglomerate together declining the available sites for electrochemical process. Compared with Cu-MnS with 1PVP, uniform and great surface adhesion on nickel foam was observed for Cu-MnS with 2PVP, which is expected have a more number of accessible charge transfer channels for redox reaction and much more effective surface due to their synergistic effect [31].

Figure 4. HR-SEM images: low magnification images of (**a**) MnS, (**c**) Cu-MnS and high-magnifications images of (**b**) MnS, (**d**) Cu-MnS.

Moreover, the 1PVP and 2PVP might have a synergistic effect and boost the electrical contact between MnS and Cu, which provides a large surface area, and might enhance the electrochemical performance. The proposed method is easier than the already published hydrothermal methods using PVP with high thermal and mechanical conductivity, which would establish the high electrical conductivity that further increases the rate performance. Interestingly, when the PVP concentration was prolonged further to 0.3 g (Figure 5f), all leaves shrank and a few degraded, which provides a crouched specific surface accessible to the aqueous electrolyte as well as enhanced electron transfer resistance compared to Cu-MnS with 1PVP and 2PVP, which should have a significant diminishing effect on the specific capacitance.

A thermogravimetric study was carried out over the temperature range of 50–800 °C; Figure S1 presents the corresponding curve (weight loss (%) versus temperature). The as-wrinkled nanotubes of MnS-7 h, containing mainly metal and sulfur precursors, degraded continuously between 100 °C to 800 °C, where a steady weight loss of 14.55% was observed, which could be due to the solvent present in the as-wrinkled nanotubes and to the evaporation of water. The agglomeration of wrinkled nanotubes (Cu-MnS) degrades via three main steps. The first weight loss of approximately 2% was observed between 100 °C to 450 °C, which might be due to the evaporation of water. A second step occurs between 450 °C to 580 °C where a steady weight loss of 20% was observed, which may be due to the decomposition of the metal precursor. The third weight loss of 2.52% until 800 °C was assigned to further decomposition of the metal precursors. Improved thermal stability was observed with respect to the pure Cu and 2PVP content in MnS up to 550 °C and a continuous weight loss was observed between 550 and 800 °C due to the decomposition of PVP.

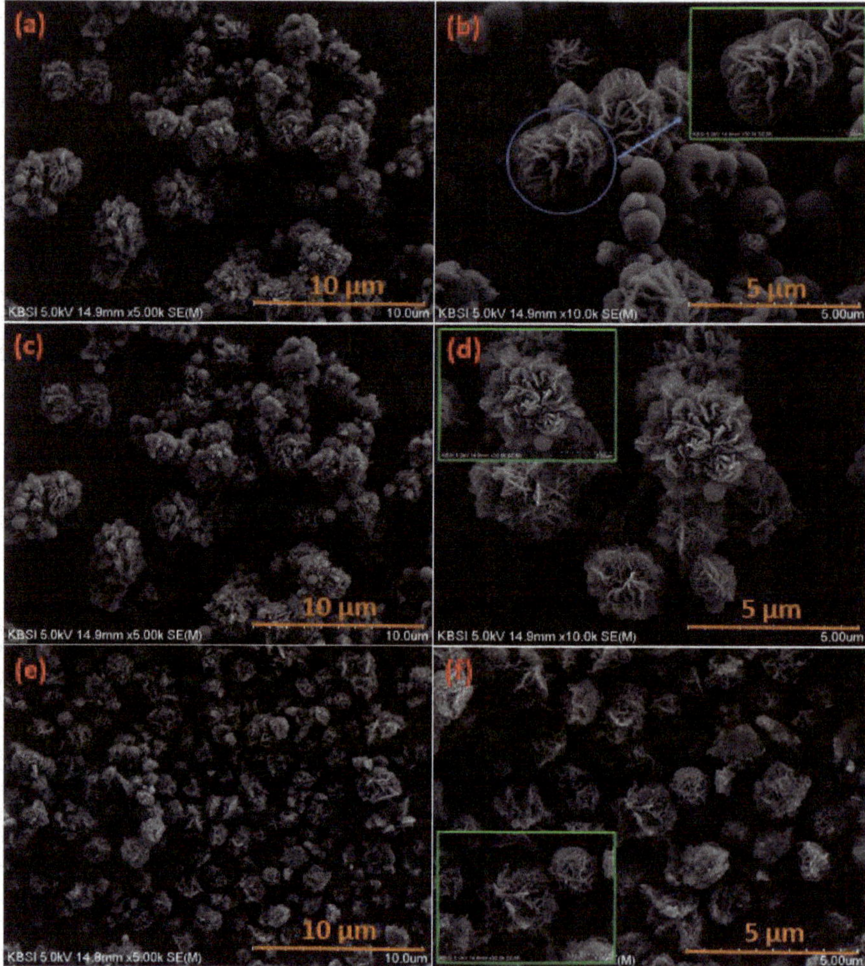

Figure 5. HR-SEM images: Cu-MnS with 1PVP, Cu-MnS with 2PVP and Cu-MnS with 3PVP (**a,c,e**) low magnification images and (**b,d,f**) high-magnifications images.

Based on the above results, all the characterizations showed that the Cu-MnS with 2PVP strongly attached to the nickel foam substrate using the brush-coated method combining a hydrothermal reaction. The assembled Cu-MnS with 2PVP electrodes displayed cabbage plant-like nanostructures with two dominances for the increase in electrochemical capacitor performance: the large surface area ameliorated from the PVP composition, which could contribute sufficient electro-active species and minimize the ion diffusion path for electrolyte ions during the electrochemical process; and high conductivity, which could affect the kinetics of electron transport [32–34].

To explore the electrochemical capacitor behavior of the as-prepared MnS, Cu-MnS, and Cu-MnS with PVP electrodes, cyclic voltammetry (CV) was performed using a three-electrode cell in 2 M KOH aqueous solution at the scan rate of 25 mV/s with a potential range from −0.5 to 0.5 V using Ag/AgCl and Pt wire as the reference and counter electrode. CV, GCD, and EIS measurements are one of the ideal tools for measuring the electrochemical performance of the electrode material for supercapacitors. As shown in Figure 6, the CV curve of the MnS electrode prepared for 7 h exhibited the largest surface area compared to the MnS-5 and MnS-10 h-based electrodes, indicating better electrochemical performance. As reported in the literature, such a high specific capacitance was attributed to the unique 1-D nanostructure of the MnS nanorods (Figure 4a) and the charge storage mechanism of the MnS nanorods is mainly a surface process, which consists of the intercalation/de-intercalation of alkali cations. The electrochemical properties including GCD, and EIS curves for MnS-5, MnS-7, and MnS-10 h are shown in supporting information (Figure S2). The specific capacitance of MnS-7 h as a function of different current density is shown in Figure S2c. The maximum specific capacitance is as high as 154.26 F g^{-1} at 1 A g^{-1}, which is higher than that of MnS-5 and MnS-10 h based electrodes. Figure S3 shows the CV curve of the bare nickel foam at scan rate of 25 mV s^{-1} in same amount of KOH electrolyte and it can be concluded that the capacitance is much smaller than that of the MnS-5, MnS-7, and MnS-10 h based electrodes, suggesting that the current collector does not contribute to the capacitance.

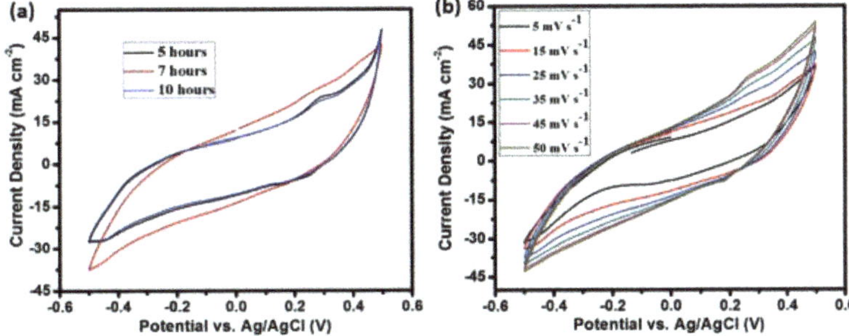

Figure 6. (a) Electrochemical performance of supercapacitors based on MnS coated on nickel foam with varying preparation times for 5, 7, and 10 h. (b) CV curves of MnS prepared for 7 h based supercapcitors for different scan rates in 2 M KOH solution.

Figure 7a,b also showed the influence of the different mass ratios of PVP in Cu-MnS on the capacitor performance. Figure 7a,b present the CV traces of Cu-MnS, Cu-MnS with the 1PVP, 2PVP, and 3PVP electrodes. All PVP-modified Cu-MnS electrodes displayed excellent pseudo-capacitor behavior. The Cu-MnS with 2PVP showed an exceptional area and higher current density values compared to the other electrodes, which indicates a rapid charge transfer rate and high capacitance at a scan rate of 5 mV s^{-1} (Figure 7b) [35]. Figure 7c shows the CV curves of the Cu-MnS with 2PVP at scan rates of 5, 15, 25, 35, 45, and 50 mV s^{-1}. These curves indicate a rapid charge transfer rate with enhanced current density. The specific capacitance (SC) decreased with increasing scan rate due to the increase in internal resistance of the electrode, produced by the crash of ionic species, which do not have sufficient time to pass electrolyte ions and electrons deep into the micropores/mesopores of the nanocomposite material [36,37]. Moreover, previous results suggested that micropores and mesopores with sizes less than ~5 nm have lower activity at high scan rate/current densities. As shown in Figure 7c and Figure S3 (CV curves of Cu-MnS, Cu-MnS with 1PVP and 3PVP with different scan rates), the positive and negative shifts observed at higher scan rates, which replicate the imperfect redox reaction and the loss of a few peaks, indicate partial loss of the redox reaction [38].

Figure 8a shows the GCD measurements for the resulting electrodes tested at a current density of 1 A g^{-1} in a 2 M KOH solution. The SC, and energy and power densities were calculated from the GCD curves using Equations (S3)–(S5) in supporting information. Inconsistent with the CV results, the PVP-based electrodes showed the longest discharge time and the calculated SC of Cu-MnS, Cu-MnS with 1PVP, 2PVPm, and 3PVP were 203.83, 251.8, 833.58, and 495.58 F g^{-1} at 1 A g^{-1}. The galvanostatic SC of the Cu-MnS with the 2PVP electrode material was performed at current densities ranging from 1 to 8 A g^{-1}. These curves appear to be almost symmetrical over the potential range of 0.4 to -0.5 V, indicating that the prepared material has greater reversibility for the charge/discharge process [39]. Therefore, the synthesized Cu-MnS with 2PVP can be considered an ultimate supercapacitor electrode material. In general, the high charge transfer rate and lower surface area of the electrode restrained ion and electron transport at the electrochemical interfaces, which led to a significant decrease in SC [36]. The introduction of copper and PVP to MnS can enhance these factors and successfully enriched the in-plane ion diffusion coefficient and SC. Importantly, the synergistic effect between Cu, PVP, and MnS helped increase the contact area between the aqueous electrolyte and electrode materials, which produced abundant electro-active sites. The highest SC of Cu-MnS with 2PVP (833.58 and 581.64 F g^{-1} at 1 and 2 A g^{-1}) in the current case indicates the superior capacitive performance compared to the previously reported studies based on MnS and their composite. The CV curves at different scan rates, the GCD curves at different currents, and Nyquist plots of bare MnS for different deposition times (5, 7, and 10 h), Cu-MnS, and Cu-MnS with 1 and 3PVP electrodes, are shown in the supporting information (Figures S2 and S4–S6). Additionally, the GCD test was performed for bare nickel foam (Figure S7) and the discharge time is very low and it can be ignored compared with Cu-MnS with 2PVP based electrodes.

EIS further revealed the precise electrical conductivity of the electrode in the frequency region from 100 kHz to 0.01 Hz under open-circuit conditions at room temperature (Figure 8c). The Nyquist plots in Figure 6f for the Cu-MnS and Cu-MnS with the PVP-based thin film electrodes showed two distinct regions in the plot—i.e., a distorted semicircle at the high frequency region and a straight line in the low frequency region—which represents the charge transfer process and capacitive nature of the electrode, respectively. Similar high frequency intercepts were observed for the Cu-MnS and Cu-MnS with 2PVP based electrodes, indicating that they have similar ohmic resistance (the combination of contact resistance between the active materials on the Ni foam and the current collector, the intrinsic resistance of the active materials, and the ionic resistance of the aqueous electrolyte). The Cu-MnS with 2PVP composite electrode had a lower charge transfer resistance (R_{ct}), which was significantly lower than that of the Cu-MnS, 1PVP, and 3PVP-based electrodes. This suggests that the Cu and appropriate amount of PVP in MnS can decrease the R_{ct}, significantly. This shows that the low Faraday resistance contributes to the enhanced SC and power of the Cu-MnS with 2PVP composite electrode in a KOH solution [40,41].

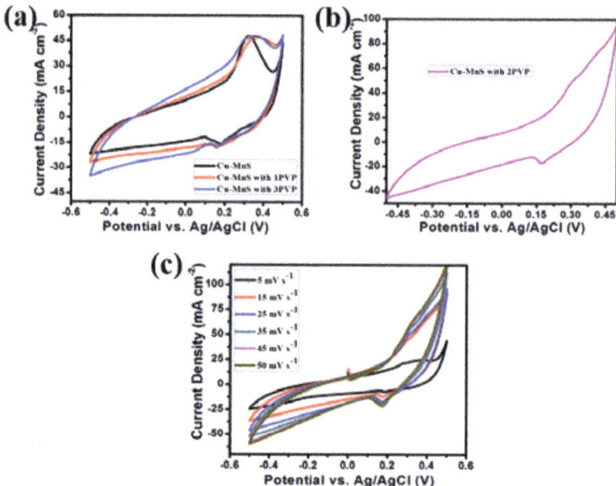

Figure 7. Cyclic voltammetry of (**a**) Cu-MnS, Cu-MnS with 1PVP, Cu-MnS with 3PVP and (**b**) Cu-MnS with 2PVP curves at scan rate of 25 mV s^{-1}. (**c**) CV of Cu-MnS with 2PVP electrodes at various scan rate from 5 to 50 mV s^{-1}.

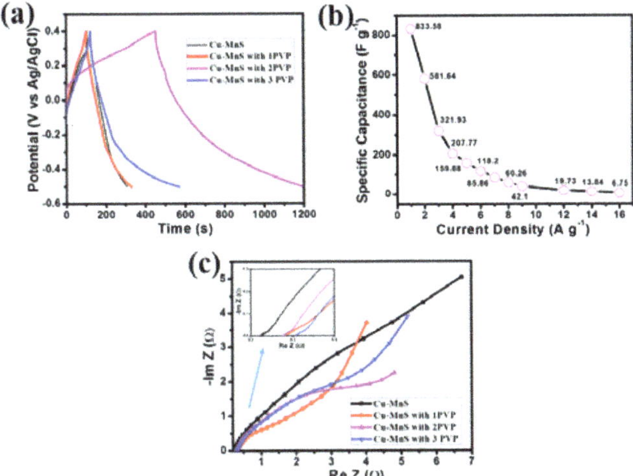

Figure 8. (a) Galvanostatic charge–discharge profile of four different electrodes at current density of 1 A g^{-1} in 2 M KOH solution and (b) specific capacitance values calculated from the GCD curves of Cu-MnS with 2PVP electrodes as a function of current densities. (c) Nyquist profile of Cu-MnS, Cu-MnS with 1PVP, Cu-MnS with 2PVP, and Cu-MnS with 3PVP electrodes respectively.

The long-term stability of the Cu-MnS and Cu-MnS with 2PVP electrodes was also performed at a current density of 1 A g^{-1} for 1000 cycles. Figure 9 shows that the specific capacitance decreased from 203.83 to 187.11 F g^{-1} during stability test after 1000 cycles for Cu-MnS electrodes, implying the deactivation of electrode in KOH. While the specific capacitance of Cu-MnS with 2PVP electrode is for higher than that of Cu-MnS and only 3.05% decrement was observed, indicating excellent cycling stability. As illustrated in the Figure S8, the surface morphology was well maintained and preserved with little structural deformation after 1000 cycles. In this study, the presence of an appropriate amount of PVP in Cu-MnS is favorable for improving the electrochemical performance of the electrode and the existence of Cu was inclined to enhance the electrical conductivity. The Cu-MnS with 2PVP electrode is a good reference for researchers to design and fabricate new electrode materials with enhanced capacitive performance.

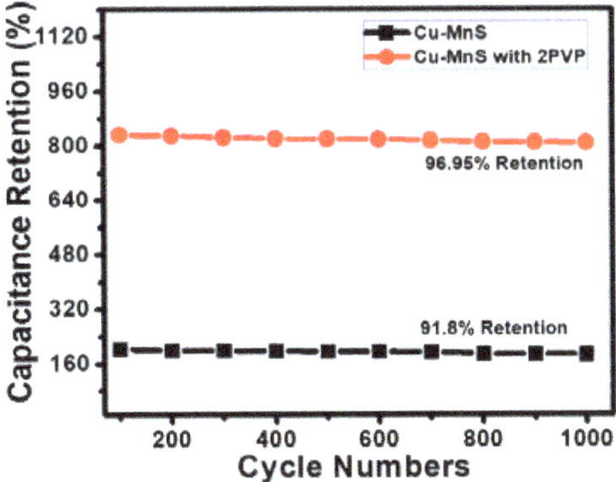

Figure 9. Capacitance retention as a function of the cycle numbers at current density of 1 A g^{-1} in 2 M KOH solution.

4. Conclusions

In summary, cabbage plant-like nanostructured Cu-MnS with 2PVP materials were synthesized by a facile and low-cost hydrothermal approach. The Cu-MnS with 2PVP samples on Ni foam were investigated as the free-standing electrode material for electrochemical supercapacitor applications and exhibited a higher specific capacitance of 833.58 F g^{-1} at 1 A g^{-1}, good rate capability and extraordinary electrochemical cycling performance. In addition, Cu-MnS and Cu-MnS with 1 and 3 PVP produced a specific capacitance of 203.83, 251.8, and 495.58 F g^{-1} at 1 A g^{-1}. The excellent electrochemical properties were attributed mainly to the following two factors: (1) the large surface area ameliorated from the PVP composition, which could contribute sufficient electro active species and minimize the ion diffusion path for electrolyte ions during the electrochemical process, electron transfer, and better accommodation to the volume change during long-term charge–discharge process; and (2) its high conductivity and surface area, which could affect the electron transport kinetics. These results suggest that the electrode material is a good candidate for high energy and power density applications that are practical for large scale use.

Supplementary Materials: The following are available online at http://www.mdpi.com/1996-1073/11/6/1590/s1, Figure S1. Thermogravimetric images of as-prepared MnS, Cu-MnS and Cu-MnS with 2PVP samples. Figure S2. (a) Galvanostatic charge–discharge curves of supercapcitors using MnS material with different preparation time. (b) GCD curves of MnS-7 h at different current densities. (c) Specific capacitance calculated from the GCD curves of MnS-7 h as a function of current densities. (d) Nyquist curves in 2 M KOH solution at frequency range from 100 kHz to 0.1 Hz. Figure S3. CV curve of bare nickel foam in KOH electrolyte at scan rate of 25 mV s^{-1}. Figure S4. Cyclic voltammetry of (a) Cu-MnS, (b) Cu-MnS with 1PVP, and (c) Cu-MnS with 3PVP with different scan rates in 2 M KOH solution. Figure S5. (a) Galvanostatic charge–discharge curves of Cu-MnS at different current densities in 2 M KOH solution. (b) Specific capacitance of Cu-MnS electrode calculated from the GCD curves as a function of current densities. Figure S6. (a) Galvanostatic charge–discharge curves of Cu-MnS with 1PVP at different current densities in 2 M KOH solution. (b) Specific capacitance of Cu-MnS electrode calculated from the GCD curves as a function of current densities. Figure S7. GCD curve of bare nickel foam in KOH electrolyte. Figure S8. HR-SEM image of Cu-MnS with 2PVP after stability test.

Author Contributions: S.S.R wrote the manuscript and carried out experiment with support from J.-W.A. and I.K.D. J.-W.A. helped during revision time with S.S.R. B.N and B.J.-S developed the theory and helped during experiment section. T.N.V.K and C.I.-H verified the experimental results. H.-J.K. investigate and supervised the funding of this work.

Funding: This research received no external funding.

Acknowledgments: This work was supported by BK 21 PLUS, Creative Human Resource Development Program for IT Convergence, Pusan National University, Busan, South Korea for its financial support. In addition, this work was supported by the Human Resources Program in Energy Technology of the Korea Institute of Energy Technology Evaluation and Planning (KETEP) granted financial resource from the Ministry of Trade, Industry & Energy, Republic of Korea. (no. 20164010200940). Finally, thanks to the KBSI for the instrumental characterizations.

Conflicts of Interest: The authors declare no conflict of interest.

References

1. Rao, S.S.; Punnoose, D.; Bae, J.H.; Durga, I.K.; Varma, C.V.T.; Naresh, B.; Subramanian, A.; Raman, V.; Kim, H.J. Preparation and electrochemical performances of NiS with PEDOT:PSS chrysanthemum petal like nanostructure for high performance supercapacitors. *Electrochim. Acta* **2017**, *254*, 269–279.

2. Rao, S.S.; Durga, I.K.; Kundakarla, N.; Punnoose, D.; Gopi, C.V.V.M.; Reddy, A.E.; Jagadeesh, M.; Kim, H.J. A hydrothermal reaction combined with a post anion-exchange reaction of hierarchically nanostructured NiCo$_2$S$_4$ for high-performance QDSSCs and supercapacitors. *New J. Chem.* **2017**, *41*, 10037–10047. [CrossRef]

3. Xiang, D.; Bote, Z.; Liang, Z.; Zongping, S. Molten salt synthesis of nitrogen-doped carbon with hierarchical pore structures for use as high-performance electrodes in supercapacitors. *Carbon* **2015**, *93*, 48–58.

4. Bote, Z.; Lei, Z.; Qiaobao, Z.; Dongchang, C.; Yong, C.; Xiang, D.; Yu, C.; Ryan, M.; Xunhui, X.; Bo, S.; et al. Rational design of nickel hydroxide-based nanocrystals on graphene for ultrafast energy storage. *Adv. Energy Mater.* **2018**, *8*, 1702247.

5. Shuge, D.; Bote, Z.; Chong, Q.; Dongchang, C.; Dai, D.; Bo, S.; Ben, M.D.; Jianwei, F.; Chenguo, H.; Ching-Ping, W.; et al. Controlled synthesis of three-phase Ni_xS_y/rGO nanoflake electrodes for hybrid supercapacitors with high energy and power density. *Nano Energy* **2017**, *33*, 522–531.

6. Bote, Z.; Dongchang, C.; Xunhui, X.; Bo, S.; Renzong, H.; Qiaobao, Z.; Benjamin, H.R.; Gordon, H.W.; Dongxing, Z.; Yong, D.; et al. A high-energy, long cycle-life hybrid supercapacitor based on graphene composite electrodes. *Energy Storage Mater.* **2017**, *7*, 32–39.

7. Gao, R.; Zhang, Q.; Soyekwo, F.; Lin, C.; Lv, R.; Qu, Y.; Chen, M.; Zhu, A.; Liu, Q. Novel amorphous nickel sulfide@CoS double-shelled polyhedral nanocages for supercapacitor electrode materials with superior electrochemical properties. *Electrochim. Acta* **2017**, *237*, 94–101. [CrossRef]

8. Huang, L.; Hou, H.; Liu, B.; Zeinu, K.; Yuan, X.; Zhu, X.; He, X.; Wu, L.; Hu, J.; Yang, J. Phase-controlled solvothermal synthesis and morphology evolution of nickel sulfide and its pseudocapacitance performance. *Ceram. Int.* **2017**, *43*, 3080–3088. [CrossRef]

9. Song, X.; Tan, L.; Wang, X.; Zhu, L.; Yi, X.; Dong, Q. Synthesis of CoS@rGO composites with excellent electrochemical performance for supercapacitors. *J. Electroanal. Chem.* **2017**, *794*, 132–138. [CrossRef]

10. Subramani, K.; Sudhan, N.; Divya, R.; Sathish, M. All-solid-state asymmetric supercapacitors based on cobalt hexacyanoferrate-derived CoS and activated carbon. *RSC Adv.* **2017**, *7*, 6648–6659. [CrossRef]

11. Adnan, Y.; Dewei, C.; Sean, L. Ethanol-directed morphological evolution of hierarchical CeOx architectures as advanced electrochemical capacitors. *J. Mater. Chem. A* **2015**, *3*, 13970–13977.

12. Zhu, G.; He, Z.; Chen, J.; Zhao, J.; Feng, X.; Ma, Y.; Fan, Q.; Wang, L.; Huang, W. Highly conductive three-dimensional MnO_2–carbon nanotube–graphene–Ni hybrid foam as a binder-free supercapacitor electrode. *Nanoscale* **2014**, *6*, 1079–1085. [CrossRef] [PubMed]

13. Wu, D.; Xu, S.; Zhang, C.; Zhu, Y.; Xiong, D.; Huang, R.; Qi, R.; Wang, L.; Chu, P.K. Three-dimensional homo-nanostructured MnO_2/nanographene membranes on a macroporous electrically conductive network for high performance supercapacitors. *J. Mater. Chem. A* **2016**, *4*, 11317–11329. [CrossRef]

14. Chhowalla, M.; Shin, H.S.; Eda, G.; Li, L.J.; Loh, K.P.; Zhang, H. The chemistry of two-dimensional layered transition metal dichalcogenide nanosheets. *Nat. Chem.* **2013**, *5*, 263–275. [CrossRef] [PubMed]

15. Sahoo, S.; Mondal, R.; Late, D.J.; Rout, C.S. Electrodeposited Nickel Cobalt Manganese based mixed sulfide nanosheets for high performance supercapacitor application. *Microporous Mesoporous Mater.* **2017**, *244*, 101–108. [CrossRef]

16. Rao, S.S.; Durga, I.K.; Gopi, C.V.V.M.; Tulasivarma, C.V.; Kim, S.K.; Kim, H.J. The effect of TiO_2 nanoflowers as a compact layer for CdS quantum-dot sensitized solar cells with improved performance. *Dalton Trans.* **2015**, *44*, 12852–12862. [CrossRef] [PubMed]

17. Rao, S.S.; Gopi, C.V.V.M.; Kim, S.K.; Son, M.K.; Jeong, M.S.; Savariraj, A.D.; Prabakar, K.; Kim, H.J. Cobalt sulfide thin film as an efficient counter electrode for dye-sensitized solar cells. *Electrochim. Acta* **2014**, *133*, 174–179.

18. Cheng, J.; Yan, H.; Lu, Y.; Qiu, K.; Hou, X.; Xu, J.; Han, L.; Liu, X.; Kim, J.K.; Luo, Y. Mesoporous $CuCo_2O_4$ nanograsses as multi-functional electrodes for supercapacitors and electro-catalysts. *J. Mater. Chem. A* **2015**, *3*, 9769–9776. [CrossRef]

19. Pang, H.; Wang, S.; Li, G.; Ma, Y.; Li, J.; Li, X.; Zhang, L.; Zhang, J.; Zheng, H. Cu superstructures fabricated using tree leaves and Cu–MnO_2 superstructures for high performance supercapacitors. *J. Mater. Chem. A* **2013**, *1*, 5053–5060. [CrossRef]

20. Cai, M.; Qian, H.; Wei, Z.; Chen, J.; Zheng, M.; Dong, Q. Polyvinyl pyrrolidone-assisted synthesis of a Fe_3O_4/graphene composite with excellent lithium storage properties. *RSC Adv.* **2014**, *4*, 6379–6382. [CrossRef]

21. Zhang, X.; Wang, J.; Jia, H.; Yin, B.; Ding, L.; Xu, Z.; Ji, Q. Polyvinyl pyrrolidone modified graphene oxide for improving the mechanical, thermal conductivity and solvent resistance properties of natural rubber. *RSC Adv.* **2016**, *6*, 54668–54678. [CrossRef]

22. Joshi, K.V.; Joshi, B.K.; Harikrishna, U.; Patel, M.B.; Menon, S.K. Polyvinyl pyrrolidone modified ZnS nanoparticles as a highly selective and sensitive nanosensor for the iodide ion. *Anal. Methods* **2013**, *5*, 4973–4977. [CrossRef]

23. Yang, G.W.; Xu, C.L.; Li, H.L. Electrodeposited nickel hydroxide on nickel foam with ultrahigh capacitance. *Chem. Commun.* **2008**, *48*, 6537–6579. [CrossRef] [PubMed]

24. Lu, M.; Yuan, X.P.; Guan, X.H.; Wang, G.S. Synthesis of nickel chalcogenide hollow spheres using an L-cysteine-assisted hydrothermal process for efficient supercapacitor electrodes. *J. Mater. Chem. A* **2017**, *5*, 3621–3627. [CrossRef]

25. Xia, Q.; Xu, M.; Xia, H.; Xie, J. Nanostructured Iron Oxide/Hydroxide-based electrode materials for supercapacitors. *ChemNanoMat* **2016**, *2*, 588–600. [CrossRef]

26. Liu, R.M.; Jian, Z.X.; Liu, Q.; Zhu, X.D.; Liu, L.; Ni, L.; Shen, C.C. Novel red blood cell shaped α-Fe$_2$O$_3$ microstructures and FeO(OH) nanorods as high capacity supercapacitors. *RSC Adv.* **2015**, *5*, 91127–91133. [CrossRef]

27. Kim, H.J.; Kim, C.W.; Punnoose, D.; Gopi, C.V.V.M.; Kim, S.K.; Prabakar, K.; Rao, S.S. Nickel doped cobalt sulfide as a high performance counter electrode for dye-sensitized solar cells. *Appl. Surf. Sci.* **2015**, *328*, 78–85. [CrossRef]

28. Ummartyotin, S.; Bunnak, N.; Juntaro, J.; Sain, M.; Manuspiya, H. Synthesis and luminescence properties of ZnS and metal (Mn, Cu)-doped-ZnS ceramic powder. *Solid State Sci.* **2012**, *14*, 299–304. [CrossRef]

29. Ramachandran, R.; Saranya, M.; Grace, A.N.; Wang, F. MnS nanocomposites based on doped graphene: Simple synthesis by a wet chemical route and improved electrochemical properties as an electrode material for supercapacitors. *RSC Adv.* **2017**, *7*, 2249–2257. [CrossRef]

30. Anothumakkool, B.; Torris, A.T.A.; Bhange, S.N.; Badiger, M.V.; Kurungot, S. Electrodeposited polyethylenedioxythiophene with infiltrated gel electrolyte interface: A close contest of an all-solid-state supercapacitor with its liquid-state counterpart. *Nanoscale* **2014**, *6*, 5944–5952. [CrossRef] [PubMed]

31. Zhiguo, Z.; Xiao, H.; Huan, L.; Hongxia, W.; Yingyuan, Z.; Tingli, M. All-solid-state flexible asymmetric supercapacitors with high energy and power densities based on NiCo$_2$S$_4$@MnS and active carbon. *J. Energy Chem.* **2017**, *26*, 1260–1266.

32. Moon, I.K.; Lee, J.; Ruoff, R.S.; Lee, H. Reduced graphene oxide by chemical graphitization. *Nat. Commun.* **2010**, *1*, 73. [CrossRef] [PubMed]

33. Choi, B.G.; Hong, J.; Hong, W.H.; Hammond, P.T.; Park, H. Facilitated ion transport in all-solid flexible supercapacitors. *ACS Nano* **2011**, *5*, 7205–7213. [CrossRef] [PubMed]

34. Lv, T.; Yao, Y.; Li, N.; Chen, T. Highly Stretchable supercapacitors based on aligned carbon nanotube/molybdenum disulfide composites. *Angew. Chem. Int. Ed.* **2016**, *55*, 9191–9195. [CrossRef] [PubMed]

35. Yang, X.; Zhao, L.; Lian, J. Arrays of hierarchical nickel sulfides/MoS$_2$ nanosheets supported on carbon nanotubes backbone as advanced anode materials for asymmetric supercapacitor. *J. Power Sources* **2017**, *343*, 373–386. [CrossRef]

36. Kumar, A.; Sager, A.; Kumar, A.; Mishra, Y.K.; Chandra, R. Performance of high energy density symmetric supercapacitor based on sputtered MnO$_2$ nanorods. *ChemistrySelect* **2016**, *1*, 3885–3891. [CrossRef]

37. Xia, H.; Zhu, D.; Luo, Z.; Yu, Y.; Shi, X.; Yuan, G.; Xie, J. Hierarchically structured Co$_3$O$_4$@Pt@MnO$_2$ nanowire arrays for high-performance supercapacitors. *Sci. Rep.* **2013**, *3*, 2978. [CrossRef] [PubMed]

38. Radhamani, A.V.; Shareef, K.M.; Rao, M.S.R. ZnO@MnO$_2$ Core–Shell nanofiber cathodes for high performance asymmetric supercapacitors. *ACS Appl. Mater. Interfaces* **2016**, *8*, 30531–30542. [CrossRef] [PubMed]

39. Cheng, T.; Zhang, Y.Z.; Yi, J.P.; Yang, L.; Zhang, J.D.; Lai, W.Y.; Huang, W. Inkjet-printed flexible, transparent and aesthetic energy storage devices based on PEDOT:PSS/Ag grid electrodes. *J. Mater. Chem. A* **2016**, *4*, 13754–13763. [CrossRef]

40. Li, W.; Wang, S.; Xin, L.; Wu, M.; Lou, X. Single-crystal β-NiS nanorod arrays with a hollowstructured Ni$_3$S$_2$ framework for supercapacitor applications. *J. Mater. Chem. A* **2016**, *4*, 7700–7709. [CrossRef]

41. Zhu, T.; Wang, Z.; Ding, S.; Chen, J.S.; Lou, X.W. Hierarchical nickel sulfide hollow spheres for high performance supercapacitors. *RSC Adv.* **2011**, *1*, 397–400. [CrossRef]

Article

Effect of Time on a Hierarchical Corn Skeleton-Like Composite of CoO@ZnO as Capacitive Electrode Material for High Specific Performance Supercapacitors

Yedluri Anil Kumar * and Hee-Je Kim

School of Electrical Engineering, Pusan National University, Busandaehak-ro 63beon-gil, Geumjeong-gu, Busan 46241, Korea; heeje@pusan.ac.kr
* Correspondence: yedluri.anil@gmail.com or yedluri.anil@pusan.ac.kr;
  Tel.: +82-10-3054-8401; Fax: +82-51-513-0212

Received: 25 October 2018; Accepted: 20 November 2018; Published: 25 November 2018

Abstract: CoO–ZnO-based composites have attracted considerable attention for the development of energy storage devices because of their multifunctional characterization and ease of integration with existing components. This paper reports the synthesis of CoO@ZnO (CZ) nanostructures on Ni foam by the chemical bath deposition (CBD) method for facile and eco-friendly supercapacitor applications. The formation of a CoO@ZnO electrode functioned with cobalt, zinc, nickel and oxygen groups was confirmed by X-ray diffraction (XRD) analysis, X-ray photoelectron spectroscopy (XPS), low and high-resolution scanning electron microscopy (SEM), and transmission electron microscopy (TEM) analysis. The as-synthesized hierarchical nanocorn skeleton-like structure of a CoO@ZnO-3h (CZ3h) electrode delivered a higher specific capacitance (C_s) of 1136 F/g at 3 A/g with outstanding cycling performance, showing 98.3% capacitance retention over 3000 cycles in an aqueous 2 M KOH electrolyte solution. This retention was significantly better than that of other prepared electrodes, such as CoO, ZnO, CoO@ZnO-1h (CZ1h), and CoO@ZnO-7h (CZ7h) (274 F/g, 383 F/g, 240 F/g and 537 F/g). This outstanding performance was attributed to the excellent surface morphology of CZ3h, which is responsible for the rapid electron/ion transfer between the electrolyte and the electrode surface area. The enhanced features of the CZ3h electrode highlight potential applications in high performance supercapacitors, solar cells, photocatalysis, and electrocatalysis.

Keywords: nanorod structure; nanocorn structure; hierarchical nanocorn skeleton-like structure; energy storage devices

1. Introduction

In recent years, with the increasing energy and power demands of the modern world, the continuous depletion of fossil energy and the continuous growth of the global economy has prompted considerable interest in the generation of renewable, clean and efficient energy (particularly in terms of the management, storage, and production of this precious energy) [1–4]. Among the electrical energy storage, ultra-capacitors have fascinated researchers and received enormous industrial attention because of their high charge–discharge current capability, high power density, very high efficiency, stable cycling performance, environmentally friendly nature and extensive temperature range compared to fuel cells [5]. Batteries have higher energy densities than traditional capacitors and conventional dielectric capacitors [6,7]. Therefore, major research has been directed towards the progression of supercapacitors with minimal sacrifice of the very high energy density and long cycling stability [8,9].

Supercapacitors are used widely in high power applications such as portable electronic devices, renewable energy storage devices, and hybrid electric vehicles [10]. Currently, the research into supercapacitors has focused mainly on the specific energy of supercapacitors and electrode materials and their micro- or nano-morphology [11]. In addition, continuous research has been concerned with designing long cycling performance electrode materials with stability, electrolyte and assembly technology [12–14]. Hence, the study of electrode materials has become a field of intense research activity [15].

Supercapacitors can be classified into two categories based on the energy storage mechanism: capacitors made typically from metal oxides or hydroxides and electric double layer capacitors (EDLCs) [16,17]. In EDLCs, carbon-based materials are commonly used in supercapacitor electrodes owing to their high power density, low cost, controllable porosity, and ease of process ability [18]. On the other hand, the relatively low energy density and volumetric capacitance has limited their efficient use for higher potential supercapacitor applications [19–21]. Some carbon-based materials, such as activated carbon and carbon nanotubes, were employed as electrode-based materials [22]. Nevertheless, in recent years, extensive attention has been paid to the surface area of the electrode [23]. In terms of capacitance, transition materials can have better capacitive performance than carbon-based EDLC electrodes, resulting from the interface, and they are thus categorized as supercapacitors. Capacitive materials contain metal oxides and sulfides, particularly metal oxide materials with higher specific capacitance, such as CoO, MnO_2, Fe_2O_3, and Co_3O_4 [24]. They have been considered promising anode materials compared to other capacitor materials and show more stable performance than carbon materials, which can result in excellent electrochemical capacitances as they can provide various oxidation states for capable faradic reactions [25]. Among those metal oxides, CoO is an ideal supercapacitor electrode material that has been shown to be promising for various applications due to its high redox reactivity, good electric conduction, low cost, high theoretic capacitance and ecofriendliness [26]. Moreover, CoO suffers from rapid capacity decay due to large volume expansion, limited ion transport kinetics and poor conductivity, cycling stability, and rate performances.

To overcome this problem, binary metal oxides show good electrochemical performance because they provide multiple redox reactions and have very high electrical conductivity. After the reaction, the two oxide interfaces formed exhibit new and interesting properties due to diffusion phenomena and proximity, which have become important in the nanoscale range. Binary metal oxide materials delivered excellent capacitances compared to single metal oxides because of their beneficial oxidation states and good electrical conductivity [27]. Among these metal oxides, ZnO has been described to be a good electrode potential material for Electrochemical capacitors (ECs) because of its simple fabrication, ecofriendliness and higher catalytic activity, as well as its high theoretical capacitance, mechanical stability, and high chemical stability. More surprisingly, ZnO/CoO exhibits good performance with superior reversible capacity and cycling stability compared to $ZnCo_2O_4$ [28]. The electrode material $CoO@ZnO$ has been associated with the synergetic effects of the different components. Moreover, the hierarchical corn skeleton possesses the most closely-packed geometry, which allows for good electrical conductivity and a large surface area suitable for a high-performance faradic reaction. It is expected that the synergistic combination of these two materials (CoO and ZoO) could pave the way to enhanced electrochemical properties. More recently, Cai et al. described $ZnO@Co_3O_4$ core/shell heterostructures using facile hydrothermal synthesis. In contrast, the performance of metal oxide based materials is still lower. Nevertheless, it is a big challenge to develop metal oxides with different nanostructures and raise the potential of flexible electrochemical capacitors [29]. It is estimated that the similar $CoO@ZnO$ should be an excellent electrode material.

In this study, CoO and ZnO electrodes were prepared and $CoO@ZnO$-1h (CZ1h), $CoO@ZnO$-3h (CZ3h) and $CoO@ZnO$-7h (CZ7h) electrodes were fabricated on Ni foam via a simple and inexpensive chemical bath deposition method (CBD) for supercapacitor applications. First, for optimization, the CoO nanorods were deposited uniformly on the as-prepared ZnO corn skeleton surface, such as CZ1h, CZ3h and CZ7h electrodes via the CBD method. The unique interconnected hierarchical corn skeleton nanostructure of the CZ3h material resulted in an improved active electrode/electrolyte interface as well as a higher active material efficiency that facilitated ion and electron transfer for electrochemical reactions. Some studies indicated that hierarchical mixed oxides, i.e., cobalt and zinc oxide nanostructures, exhibit a large surface area as well as better electrical conductivity by producing hydroxyl groups and oxygen vacancies for the faradic reaction [30]. Under optimized conditions, the as-prepared CZ1h and CZ7h electrodes exhibited specific capacitance (240 F/g and 537 F/g at 3 A/g, respectively). On the other hand, the nanorod structure of the CoO electrode and corn skeleton-like ZnO electrode exhibited good specific capacitance (274 F/g and 383 F/g at 3 A/g, respectively). The hierarchical corn skeleton of the CZ3h electrode delivered a high specific capacitance of 1136 F/g at 3 A/g and a considerably long cycling lifetime, showing 98.3% capacitance retention after 3000 cycles. In this study, electrochemical measurements indicated that the hierarchical corn skeleton of CZ3h has remarkable electrochemical capacitance with good electrical conductivity and long lifetime cycling stability. The methodologies, through well-designed combinations and synthesized methods executed in this work, are applicable for the development of energy storage devices with ideal supercapacitor behavior and extraordinary electrochemical performance.

2. Experimental Methods

2.1. Chemicals and Materials

All of the reagents used in this study were of analytical grade and used without further purification. Analytical grade chemical reagents, zinc nitrate ($Zn(NO_3)_2$ (Chemical Abstracts Service (CAS) Number 13778-30-8)), hexamethylenetetramine ($(CH_2)_6N_4$ (Chemical Abstracts Service (CAS) Number 100-97-0)), cobalt acetate ($C_4H_6CoO_4$ (Chemical Abstracts Service (CAS) Number 71-48-7)), urea (CH_4N_2O(Chemical Abstracts Service (CAS) Number 57-13-6)), acetic acid (CH_3CoOH (Chemical Abstracts Service (CAS) Number 64-19-7)), and potassium hydroxide (KOH (Chemical Abstracts Service (CAS) Number 1310-58-3)) were purchased from Sigma Aldrich chemicals.

2.2. Preparation of ZnO and CoO on Nickel Foam

In a typical synthesis, pieces of Ni foam were cleaned carefully with a 1 M HCl solution for 30 min and dialyzed for a weak to remove acids and ions. This was followed by a rinse in acetone, absolute ethanol, and deionized (DI) water in an ultrasonic bath for 20 min. A 50 mL precursor solution was prepared with 0.03 M of zinc nitrate and 0.04 M of hexamethylenetetramine (HMTA) with magnetic stirring for 30 min. The resulting light pink solution was transferred to a 50 mL beaker and the cleaned Ni foam substrates were immersed vertically into the reaction solution. Subsequently, the beaker was kept in an oven at 100 °C for 20 h. The loaded nickel foam precursor was washed carefully with absolute ethanol and DI water and kept in an oven at 100 °C for 1 h. Finally, the obtained product annealing the precursor at 350 °C for 3 h in a furnace to improve the crystallinity. A pure CoO electrode was also prepared using the same experimental method with the relevant chemicals without the addition of ZnO.

2.3. Preparation of CZ Composite Nanostructure

To synthesize CZ samples, 0.05 M cobalt acetate, 0.4 M acetic acid, and 0.04 M urea were dispersed in 50 mL of DI water with magnetic stirring for 30 min. The pink solution was transformed to a 50 mL beaker and the as-prepared ZnO electrode of Ni foam was immersed vertically into the solution and then kept in an electric oven at 100 °C for optimization of various timings, such as 1 h, 3 h and 7 h, to obtain the active electrodes. The reaction was allowed to cool to room temperature. The coating on the Ni foam was removed from the oven and rinsed carefully with absolute ethanol and DI water and maintained at 100 °C for 6 h in an oven.

2.4. Characterization

The products of the crystal structure and phase purity were characterized by powder X-ray diffraction (XRD, D8 ADVANCE, Tokyo, Japan). The morphology and microstructure of the synthesized products and elemental maps were characterized by scanning electron microscopy (SEM, s-2400, Hitachi, Tokyo, Japan) equipped with energy-dispersive X-ray spectroscopy (EDX). Amplitude-contrast transmission electron microscopy (TEM) was used to obtain information on the surface morphology and size. X-ray photoelectron spectroscopy (XPS) by VG scientific ESCALAB250 was used to study the elemental arrangement of the prepared materials.

2.5. Electrochemical Characterization

Cyclic voltammetry (CV) was carried out using a BioLogiSP150 electrochemical workstation instrument at various scan rates that ranged from 2 mVs^{-1} to 6 mVs^{-1} at a potential window of 0 V to 0.4 V. The galvanostatic charge–discharge (GCD) tests were conducted using chronopotentiometry at current densities from 3 A/g to 7 A/g at potentials ranging from 0 V to 0.4 V. The three-electrode cells were tested in a 2 M KOH aqueous solution as the electrolyte. The CZ electrodes on nickel foam were investigated directly as the working electrode, while a saturated Ag/AgCl electrode and a Pt wire were used as the reference and counter electrodes, respectively. Electrochemical impedance spectroscopy (EIS) measurements were taken over a frequency region between 0.100 Hz to 100 KHz at the open circuit potential with an alternating current (AC) potential amplitude of 5 mV. The energy density (E, W h kg^{-1}), power density (P, W kg^{-1}), and specific capacitance (C_s, Fg^{-1})) were measured from the GCD curves using the equations:

$$E = \frac{C_s \times \left(\Delta V^2\right)}{2} \tag{1}$$

$$P = \frac{E}{t} \tag{2}$$

$$C_s = \frac{I \times t}{m \times V} \tag{3}$$

where I is the discharge current, t is the discharge time, V is the potential window and m is the mass of the active electrode, respectively.

3. Results and Discussion

The growth procedures of CoO, ZnO, CZ1h, CZ3h and CZ7h electrodes on Ni foam are illustrated in Figure 1.

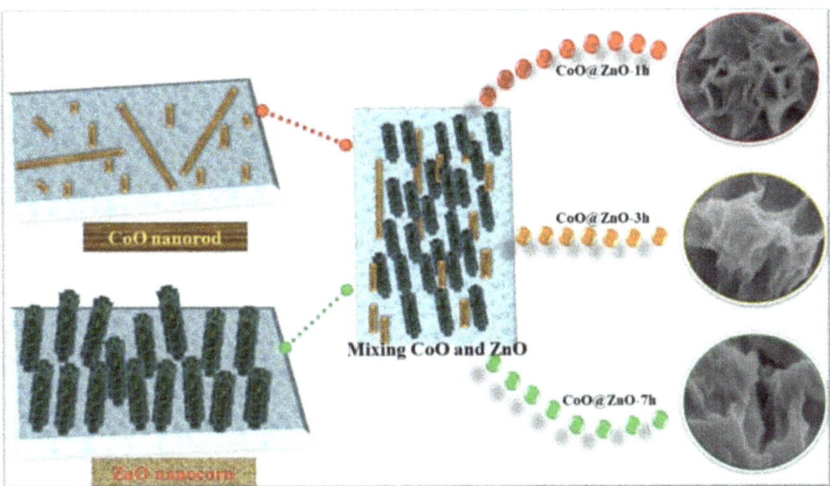

Figure 1. Illustration of the synthesis process of various nanostructures of CoO and ZnO and the effect of time on CZ composite electrodes on the Ni foam.

Characterization and Electrochemical Performance of Active Electrodes

The surface morphology, size and surface structures of the as-prepared electrodes, such as CoO and ZnO, were examined by low and high magnification Field emission scanning electron microscopy (FE-SEM) and Energy-dispersive x-ray spectrometer (EDS), as shown in Figure 2a–h.

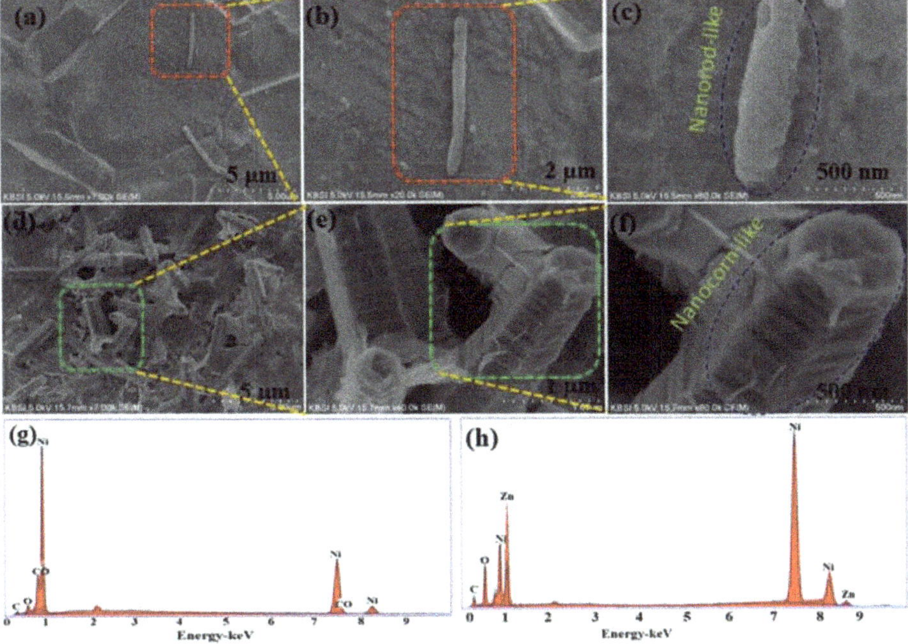

Figure 2. Low and high-magnification Field emission-scanning electron microscopy (SEM) images of the (**a**–**c**) CoO nanorod structures and (**d**–**f**) ZnO nanocorn structures and the energy-dispersive X-ray spectroscopy (EDX) pattern of CoO and ZnO (**g**,**h**) nanostructures on Ni foam.

Figure 2a–c presents SEM images of the CoO electrode on Ni foam by chemical bath deposition, indicating immature nanorods with different sizes and morphologies. Figure 2a,b shows low-resolution images of the CoO electrode. Different sized particles and the irregular morphology of the nanorod-like structure resulted in a smaller surface area of the electrode due to rapid nucleation. Figure 2c shows a high-resolution image of the CoO electrode showing a clear nanorod structure, indicating poor adhesion and a small surface area. Figure 2d–f presents low-resolution images of ZnO. The ZnO electrode showed well preserved nanocorn morphology on the Ni foam and may expose a larger surface area for growth of the active electrode, as shown in Figure 2d,e. The high-resolution SEM image of the ZnO electrode revealed the unique structure of a nanocorn morphology providing suitable diffusion channels for interactions with the electrolyte for the charge storage mechanism on the electrode surface (Figure 2f). Moreover, Figure 2f shows a high magnification of ZnO to illustrate the well-ordered nanocorn morphology on nickel foam. The high magnification SEM images (Figure 2f) show that ZnO consists of well-defined interconnected corn-shaped subunits with a size of around 200–500 nm and a thickness of around 20–30 nm. In addition, a large number of small pores with interconnected corn-like nanocorn structures were also observed. This type of unique interconnected nanocorn morphology provides suitable diffusion channels for the transport of the electrolyte and ensures efficient contact between the working electrode and electrolyte. Furthermore, EDX of CoO and ZnO (Figure 2g,h) confirmed the presence of Co, Zn, Ni, and O as the principal elemental components with no other impurities being obtained. EDX spectra of CZ1h and CZ7h (Figure S1a,b in the supporting information) confirmed the presence of Co, Zn, Ni and O as principal elemental components with no other impurities being obtained.

CZ was synthesized using the CBD method for various reaction times at 1 h, 3 h, and 7 h on Ni foam, such as CZ1h, CZ3h and CZ7h. Figure 3a–i shows low and high magnification FE-SEM images of the as-prepared samples. Figure 3a–c presents the CZ1h synthesized electrode on Ni foam. From Figure 3a,b, high magnification images of the active electrode revealed a nano-flake like structure that was formed by irregular connections with a low surface area and large pore sizes. In particular, Figure 3c indicates large pore sizes and less adhesion, making it difficult for ions or electrons to be transported within the electrode. Figure 3d–f presents the CZ3h synthesized electrode, indicating a well interconnected structure with a high surface area converted from the CZ1h electrode by the increased time. The nanostructure of CZ3h was comprised of CoO nanorods covered uniformly with the ZnO nanocorn surface area (Figure 3d,e). The high magnification SEM images of CZ3h confirmed that the synthesized composite has a non-uniform size which may be a result of the combination of particles with a corn skeleton-shape and the growth of unequal crystalline grains for the period of synthesis. Moreover, the CZ3h hierarachical nanocorn skeleton-like composites were closely connected to each other, which increased the number of electroactive sites for electrochemical reactions and formed a large surface area that can supply fast paths for the insertion and extraction of electrolyte ions, which is beneficial to the faraday reaction between the active material and electrolyte ions. Figure 3g–i shows the effects of increasing the time on the CZ7h electrode. As the annealing time was extended from 3 h to 7 h, the similar mesoporous structure of CZ7h was converted completely to disturbed nanoparticles with a weakened structure and large pores, indicating interrupted adjacent pores and a reduced surface area of the active electrode, as shown in Figure 3g,h. Figure 3i shows that the nanostructure is dispersed unevenly with voids among the particles on the surface of the active electrode. The corresponding chemical reactions occurred during the hydrothermal synthesis and are expressed using the following formulas:

$$CO(NH_2)_2 + H_2O \rightarrow 2NH_3 + CO_2 \tag{4}$$

$$H_2O + NH_3 \rightarrow OH^- + NH^{4+} \tag{5}$$

$$ZnF^+ + OH^- \rightarrow Zn(OH)F \tag{6}$$

$$CO_2 + H_2O \rightarrow CO_3{}^{2-} + 2H^+ \tag{7}$$

$$CoF^+ + xOH^- + 0.5(2-x)CO_3^{2-} + nH_2O \rightarrow Co(OH)_x(CO_3)_{0.5(2-x)} \cdot nH_2O + F^- \quad (8)$$

$$Zn(OH)F \rightarrow ZnO + HF\uparrow \quad (9)$$

$$Co(OH)_x(CO_3)_{0.5(2-x)} \cdot nH_2O + O_2 \rightarrow Co_3O_4 + CO_2\uparrow + H_2O\uparrow \quad (10)$$

The morphological features and structural information of the as-obtained samples of CoO, ZnO, and CZ1h, CZ3h and CZ7h were studied by transmission electron microscopy (TEM and High resolution-TEM), as shown in Figure S2a–f in the Supporting Information. Figure S2a–c in the Supporting Information shows a TEM image of the CoO electrode, which exhibits small sized nanorods and nanoparticle-like morphology. As shown in Figure S2a,b in the Supporting Information, irregular CoO nanoparticles were formed, resulting in a low surface area. Figure S2c in the Supporting Information shows thin CoO nanorods. Figure S2d–f in the Supporting Information presents low and high magnification TEM images of the ZnO nanocorn, showing good crystallinity with the observations from the SEM images. The low magnification TEM images of ZnO clearly showed the nanocorn nature by the nanosheet-like morphology with curling folded edges and obvious features, as shown in Figure S2d,e in the Supporting Information. Moreover, the obvious contrast shows (Figure S2f in the Supporting Information) black part and white part ligaments and indicates the mesoporous characteristics of the ZnO nanocorn-like structure. This is well consistent with the TEM observations, which were highly consistent with the SEM analysis.

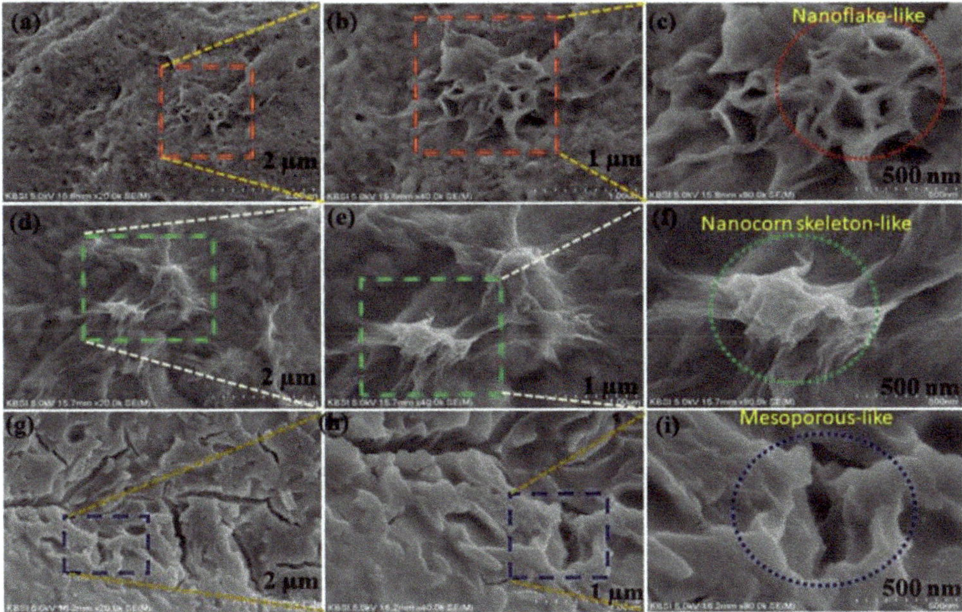

Figure 3. (**a–c**) SEM and FE-SEM images of CZ1h, (**d–f**) SEM and FE-SEM images of CZ3h and SEM and (**g–i**) FE-SEM images of CZ7h nanostructures.

Figure 4a presents low magnification images of the active electrode nanosheet-like structures. The CoO nanorods were evidently coated with ZnO corn straw nanostructures and formed a clear image, as shown in Figure 4b,c, indicating that this structure may shorten the ion diffusion length markedly for electrochemical reaction. On the other hand, EDS was performed on the CZ3h sample, as shown in Figure 4d, with active electrodes composed of Co, Zn, Ni, and O elements with no other impurities being obtained. Figure 4e,f presents different magnification images of the CZ1h electrode. The middle of the small thin layer sheet can be seen in Figure 4e, whereas the presence of CoO nanoparticles on the edge of the ZnO nanosheet was observed clearly in Figure 4f. Figure 4g,h shows the TEM images of CZ7h at two different magnifications. Figure 4g shows a thin nanosheet, supporting the corresponding SEM images, which is also seen clearly in Figure 3h. Scanning transmission electron microscopy-energy dispersive x-ray spectroscopy (STEM-EDS) mapping has been used to investigate the elemental distribution in the CZ composite. Figure 4i,j shows the STEM image of the CZ3h hierarchical corn skeleton composite and its colored elemental mapping of Co and Zn, respectively. This result clearly proved the homogeneous distribution of Co and Zn in the CZ3h corn skeleton-like composite. Therefore, this hierarchical nanocorn skeleton-like (CZ3h) structure is beneficial for efficient and fast ion and electron transport to the surface of the active materials, which enhances the effective utilization of the active material.

The crystalline phase purity and composition of the as-prepared samples (CoO, ZnO, and CZ1h, CZ3h, and CZ7h) were examined by XRD as shown in Figure 5. In the XRD pattern of CoO, the XRD peaks at $21.8°$, $36.84°$, $37.8°$, $43.3°$, $44.81°$, $49.7°$, $63.29°$, $75.2°$ and $78.40°$ 2θ were attributed to the (101), (003), (311), (400), (113), (400), (200), (622), and (311) planes, respectively. In CoO, all the XRD peaks were indexed using (Joint committee on powder diffraction standards (JCPDS) No. 48-1719) and related to the cubic phase. On the other hand, in the ZnO sample, the XRD peaks at $47.72°$, $44.8°$, $56.95°$, $64.4°$, and $68.2°$ 2θ were assigned to the (102), (400), (422), (220), and (112) planes of a well-crystallized hexagonal structure ZnO (JCPDS No. 36-1451). In the remaining samples of CZ1h, CZ3h, and CZ7h, the XRD peaks at $36.26°$, $45°$, $63.29°$, $76.20°$ and $78.40°$ 2θ were assigned to the (311), (400), (511), (622), and (622) planes of a cubic spinal structure (JCPDS No. 23-1390), respectively. Moreover, the XRD pattern in Figure 5 confirmed the successful preparation of CoO and ZnO on Ni foam. No other peaks for impurities were observed, which confirms that the cobalt oxide precursor and zinc oxide precursor had been transformed completely to the CZ composite after chemical bath deposition at $300\ °C$. In addition to these peaks, one strong peak in all samples at $51.9°$ 2θ was assigned to Ni foam. The specific surface area and pore size distribution are important factors for electroactive materials in supercapacitors. In order to evaluate the surface area properties of the as-synthesized CoO, ZnO, CZ1h, CZ3h and CZ7h electrodes, N_2 adsorption–desorption isotherms of "Brunauer–Emmett–Teller (BET) analysis" are shown in Figure 5b,c. Uniform distribution of CZ3h ensures a higher electrochemically active surface area and hence enhances the electrical conductivity. After analysis, CoO, ZnO, CZ1h and CZ7h were found to have minimal surface area and subsequently lower specific capacitance than CZ3h. The BET surface area and mean pore diameter of CoO, ZnO, CZ1h and CZ7h were found to be $22.6\ m^2\ g^{-1}$, $32.5\ m^2\ g^{-1}$, $27.3\ m^2\ g^{-1}$, $43.6\ m^2\ g^{-1}$ with an average pore size mostly below 10 nm, whereas CZ3h showed approximately $86.7\ m^2\ g^{-1}$ with an average pore size mostly below 20 nm. In addition, CZ3h showed enhanced pore size distribution and the pores were in the macroporous range. This mesoporous structure of the CZ3h electrode could enhance the specific surface area and the reaction sites at the interface of the electrolyte and electrode significantly.

XPS was also carried out to characterize the surface chemical composition and oxidation states of the chemical bonding of Co, Zn, Ni, and O on the surface of CoO, ZnO, and CZ1h, CZ3h, and CZ7h, and the results are presented in Figure 6. Figure 6b shows strong peaks for Zn 2p, Co 2p, Ni 2p and O 1s in the CZ3h survey spectrum. The high resolution Zn 2p spectrum shows the characteristic Zn $2p_{3/2}$ and Zn $2p_{1/2}$ peaks at 1022 eV and 1045 eV, respectively, confirming the presence of divalent Zn, as shown in Figure 6c. The strong peaks for Co 2p were typically deconvoluted into two peaks at 781.5 eV and 796.5 for Co $2p_{3/2}$ and Co $2p_{1/2}$, respectively, and two weak shake-up satellite peaks at 786.2 eV and 802.6 eV, as shown in Figure 6d [31,32]. The energy gap between the main peaks was approximately 15 eV, suggesting the presence of Co^{2+} and Co^{3+}. An energy gap between the main peaks and satellite peaks of around 6 eV would indicate divalent Co. Figure 6e shows the strong resolution X-ray photoelectron spectroscopy (XPS) of Ni 2p, which is typically separated into two strong peaks at 857.0 eV and 875.0 eV for Ni $2p_{3/2}$ and Ni $2p_{1/2}$, respectively, and its satellite peaks appear at 861.7 eV and 880.0 eV [33]. In addition, Figure 6f shows the O 1s spectrum with two deconvoluted peaks at 531.0 eV and 529.8 eV, suggesting hydroxide ions and metal oxygen bonds in the sample [34–36].

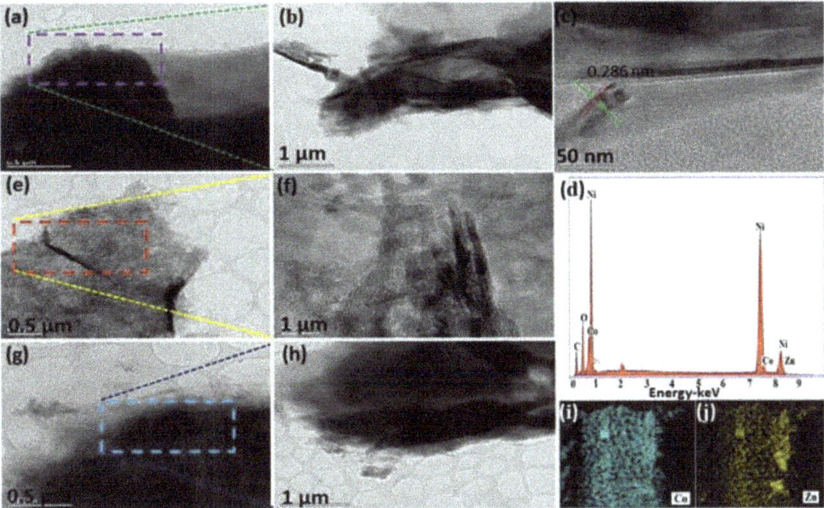

Figure 4. Transmission electron microscopy (TEM) and High resolution (HR-TEM) images for CZ3h, CZ1h and CZ7h composites. (**a–c**) presents low and high magnification TEM images of the CZ3h composite electrode and (**d**) Energy dispersive x-ray spectroscopy (EDS) spectrum, (**e,f**) CZ1h composite electrode, (**g,h**) CZ7h composite electrode and (**i,j**) elemental mapping images of Co and Zn.

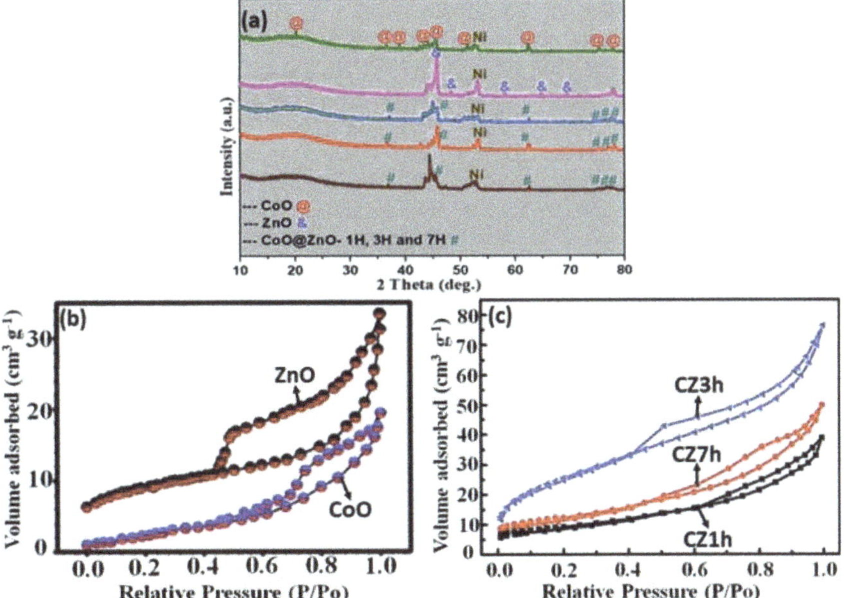

Figure 5. X-ray diffraction patterns (XRD) of all the samples. The Brunauer–Emmett–Teller (BET) surface area measured from nitrogen adsorption–desorption isotherms of (**b**) CoO and ZnO (**c**) CZ1h, CZ3h and CZ7h on Ni foam.

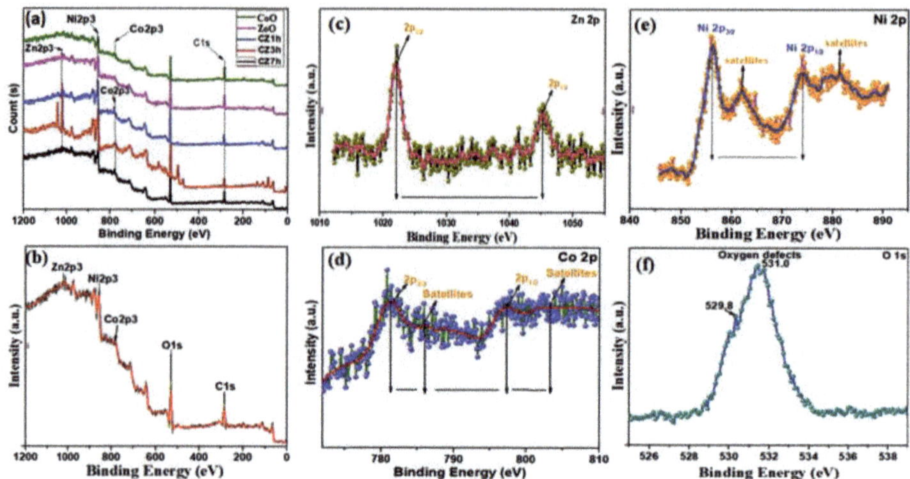

Figure 6. Shows the X-ray photoelectron spectroscopy (XPS) of (**a**) all samples and XPS spectra of CZ3h of (**b**) survey spectrum, (**c**) Zn 2p, (**d**) Co2p, (**e**) Ni 2p and (**f**) O 1s.

To examine the electrochemical behavior and supercapacitor properties of the as-prepared CoO, ZnO, and composites, the CZ1h, CZ3h, and CZ7h synthesized electrodes were first investigated in a three-electrode system (Figure 7). Figure 7a compares the CV curves of all the CoO, ZnO, CZ1h, CZ3h, and CZ7h electrodes in a 2 M KOH solution at 50 mV s^{-1} within a voltage window of 0 V to 0.4 V. The electrochemical performance of the CZ3h electrode was significantly higher than the remaining electrodes (CoO, ZnO, CZ1h, and CZ7h). The nanostructure of the CZ3h composite material shows great adhesion on the nickel foam substrate and a large surface area which produces more active sites for faradaic redox reactions and electron transport pathways, resulting in greatly enhanced electrochemical performance. Figure 7b shows the galvanostatic charge–discharge curves of all electrodes over 0 V to 0.4 V at 3 A/g. The prepared electrodes underwent the faradic redox reversible reactions. Remarkably, the CZ3h electrode exhibited a high specific capacitance, which is much larger than those for the CoO, ZnO, CZ1h, and CZ7h electrodes. The hierarchical corn skeleton structure of the CZ3h nanocomposite that was synthesized via a facile chemical bath deposition offers many remarkable advantages, such as efficient electrical contact between the nickel foam substrate and the electroactive materials and simplified electrode preparation, which act to improve capability and capacitance. Figure 7c shows the specific capacitance of the as-prepared samples calculated from the GCD curves at different current densities. For comparison, the specific capacitance of the CZ3h electrode exhibited 1136 F/g at 3 A/g, which is higher than the other electrodes. The strong synergistic effect of the CZ3h composite may contribute to the remarkable superior electrical performance compared with other electrodes such as CoO, ZnO, CZ1h and CZ7h. Figure 7d shows the CV profile of the individual CZ1h electrode. The shape of the CV curve remained the same even at a low scan rate of 2 mV s^{-1} up to a high scan rate of 100 mV s^{-1} and the GCD curves also maintained the same shape from current densities of 3 A/g to 7 A/g and exhibited a specific capacitance values of 240 F/g, 228 F/g, 214 F/g, 197 F/g and 190 F/g as shown in Figure 7e. The CV curves of the CZ3h composite electrode were investigated from 2–100 mV s^{-1} (Figure 7f). The CZ3h electrode showed a nanocorn morphology with an interconnected surface area, which is favorable for easy transportation in the electrochemical reaction. On the other hand, the GCD curves of the CZ3h electrode at various current densities (3 A/g to 7 A/g) exhibited a specific capacitance value of 1137 F/g, 1085 F/g, 892 F/g, 770 F/g and 675 F/g and are shown in Figure 7g. Construction of corn-skeleton structures with the combination of two different materials were shown to be a promising strategy to boost the greater electrochemical performance of metal oxides. Figure 7h shows the CV curves of the CZ7h electrode at various scan rates. The current density increased with increasing scan rate, and all curves showed a unique shape. The GCD curves also maintained the same shape from current density values of 3 A/g to 7 A/g and exhibited specific capacitance values of 537 F/g, 512 F/g, 502 F/g, 492 F/g and 475 F/g, as shown in Figure 7i.

Figure S3a in the Supporting Information presents the CV profile of the CoO electrode, indicating the clear shape of the CV curve from 2–100 mV s^{-1} and GCD curves also maintained the same shape from current densities of 3 A/g to 7 A/g and exhibited specific capacitance values of 274 F/g, 255 F/g, 240 F/g, 217 F/g and 203 F/g (Figure S3b, Supporting Information), respectively. Figure S3c in the Supporting Information shows the CV curves of the ZnO electrode recorded at scan rates of 2–100 mV s^{-1} within the potential range from 0 to 0.4 V. The shape of the CV curve was the same even at a high scan rate of 100 mV s^{-1}. Figure S3d in the Supporting Information shows the charge–discharge curves at current densities (3 A/g to 7 A/g) and exhibited specific capacitance values of 383 F/g, 359 F/g, 312 F/g, 281 F/g and 273 F/g, respectively.

Ion diffusion and electron transfer in the electrode materials were examined. Figure 8a presents the Nyquist plot of all the electrodes over the frequency range of 0.100 Hz to 100 KHz. Figure 8a shows that the CZ3h electrode had a small semicircle in the high frequency range and a more vertical line in the low frequency range than the CoO, ZnO, CZ1h, and CZ7h electrodes, suggesting higher charge transfer between the active electrodes and the electrolyte. These very small circles are associated with rapid ion diffusion in the CZ3h electrode. These results show that combinations of the KOH aqueous electrolyte and the intrinsic resistance of the active electrode resulted in remarkable electrochemical performance of the CZ3h electrode.

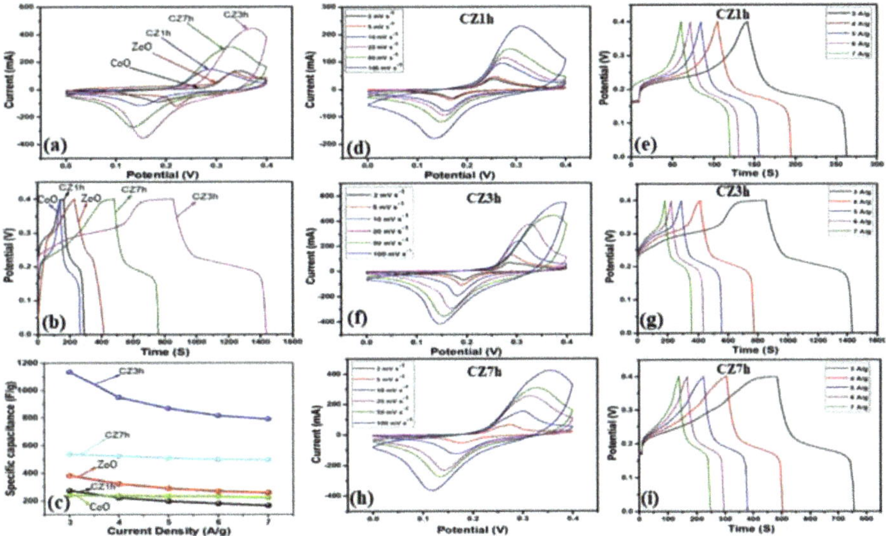

Figure 7. (a) Comparative cyclic voltammetry (CV) curves of the CoO, ZnO, CZ1h, CZ3h and CZ7h electrodes at a scan rate of 50 mV s^{-1}. (b) Comparative galvanostatic charge–discharge (GCD) curves of the CoO, ZnO, CZ1h, CZ3h and CZ7h electrodes at a current density of 3 A/g, (c) Specific capacitance vs. current density curve of the as-prepared electrodes, (d,e) CV curves and GCD curves of the CZ1h electrodes at various scan rates and current densities. (f,g) CV curves and GCD curves of the CZ3h electrodes at various scan rates and current densities. (h,i) CV curves and GCD curves of the CZ7h electrodes at various scan rates and current densities.

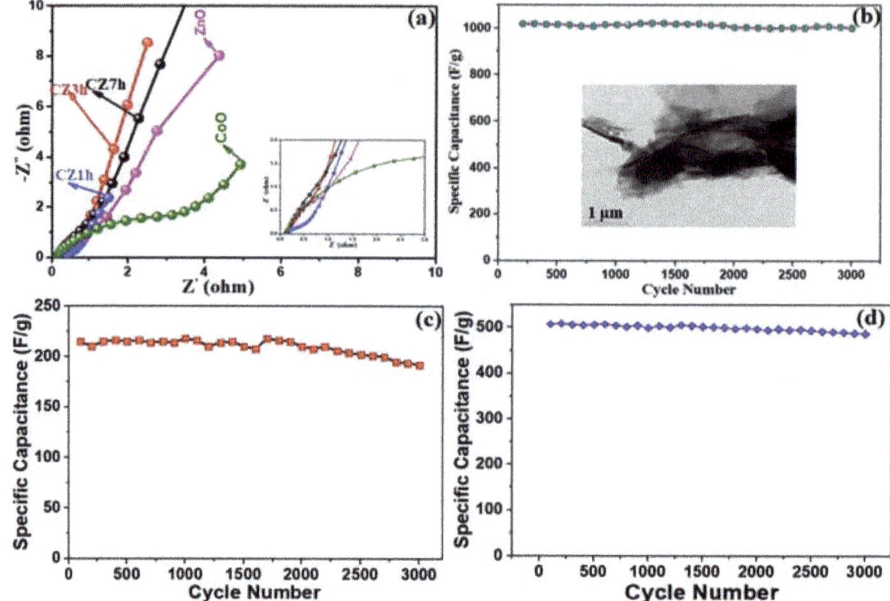

Figure 8. (**a**) Electrochemical impedance spectra (EIS) plots of the CoO, ZnO, CZ1h, CZ3h and CZ7h electrodes and (**b–d**) cycling stability performance of the CZ3h, CZ1h and CZ7h electrodes at 4 A/g for 3000 cycles.

A long-term cycle test was performed to examine the stability, which is an indispensable factor when determining the practical applicability of the CZ3h, CZ1h and CZ7h electrodes, as shown in Figure 8b–d. The cycling stability of the CZ3h, CZ1h and CZ7h electrodes were measured up to 3000 cycles by recording the GCD at a current density of 4 A/g in a three-electrode system. Figure 8b–d shows that the CZ3h electrode exhibits excellent capacitance that still retained 98.3% of the initial value after 3000 cycles, which was remarkably higher than those of CZ1h (83% initial value after 3000 cycles) and CZ7h (90.6% of the initial capacitance after 3000 cycles). This was attributed to the gradual penetration of the electrolyte into the interior of the electrode materials during cycling and the high structural stability of the electrode. Efficient electrical contact between the nickel foam substrate and the electroactive materials simplified electrode preparation, resulting in easy diffusion of the electrolyte and short ion diffusion pathways, thus leading to improved cycling stability and rate capability. To the best of our knowledge, these obtained output values for the CZ3h nanocomposite are the one of the greatest values and have good potential for supercapacitor applications compared to our previous reports and other composite materials (see Table 1).

Table 1. Comparison of various zinc- or cobalt oxide-based electrode materials and their electrochemical performances taken from recently published reports and the present work.

Electrode Materials	Electrolyte	Specific Capacitance at Current Density	Capacitance Retention (No. of Cycles)	Year, Ref.
Co_3O_4/ZnO	6 M KOH	415 F/g	93.2% (1000)	2016 (ref. [37])
$ZnCo_2O_4$	6 M KOH	542.5 F/g	95.5% (2000)	2017 (ref. [38])
$ZnCo_2O_4$	2 M KOH	776.2 F/g	84.3% (1500)	2017 (ref. [39])
Co_3O_4 nanotube	6 M KOH	574 F/g	95% (1000)	2010 (ref. [40])
CNO–ZnO//ZnO	3 M KOH	125 F/g	92% (2000)	2017 (ref. [41])
ZnO nanoflakes	2 M KOH	260 F/g	73.33% (1000)	2018 (ref. [42])
Co_3O_4 nanorods	6 M KOH	456 F/g	65% (500)	2009 (ref. [43])
$ZnCo_2O_4$	2 M KOH	647.1 F/g	91.5% (2000)	2015 (ref. [44])
C@NiCo2O4	6 M KOH	404 F/g	87.1% (1000)	2018 (ref. [45])
Co_3O_4/Graphene	6 M KOH	328 F/g	97.9% (2000)	2017 (ref. [46])
CZ3h	2 M KOH	1136 F/g	98.3% (3000)	This work

4. Conclusions

CoO, ZnO, CZ1h, CZ3h and CZ7h nanostructures were synthesized on Ni foam by an inexpensive and ecofriendly CBD method. Ni foam provides a higher surface area, excellent electrical conductivity, very high porosity, and sufficient functional groups to the material to enhance the electrochemical performance. Various electrochemical techniques were carried out to characterize the nanostructure, morphology, and structural properties of the working electrodes. The hierarchical interconnected corn skeleton of the CZ3h electrode allowed the electrolyte to penetrate directly, indicating a reduction in the ionic transfer resistance and thereby improving the working material surface area efficiency and electron transport. In a three-electrode system, the CZ3h active electrode exhibited a good specific capacitance (1136 F/g at a current density of 3 A/g). In contrast, the remaining electrodes (CoO, ZnO, CZ1h, and CZ7h) exhibited 274 F/g, 383 F/g, 240 F/g and 537 F/g at a current density of 3 A/g in the three-electrode system. Considering the encouraging electrochemical properties of the active electrodes device, it is strongly believed that hierarchical corn skeleton nanostructures will form part of new strategies for next generation energy storage devices.

Supplementary Materials: The following are available online at http://www.mdpi.com/1996-1073/11/12/3285/s1, Figure S1: EDX patterns of CZ1h and CZ7h electrodes (a,b) nanostructures on Ni foam, Figure S2: Shows the (a–c) TEM and HR-TEM images of CoO sample structures and (d–f) TEM and HR-TEM images of ZnO sample structures, Figure S3: (a,b) CV curves of the CoO and ZnO electrodes at various scan rates, (c,d) GCD curves of the CoO and ZnO electrodes at different current densities from 3 A/g to 7 A/g, Figure S4: Shows the X-ray photoelectron spectroscopy (XPS) of CZ1h and CZ7h electrodes of survey spectrums (a,b).

Author Contributions: Conceptualization, A.K.Y.; Formal analysis, H.J.K.; Investigation, A.K.Y. and H.J.K.; Methodology, A.K.Y.; Supervision, A.K.Y.; Writing–original draft, A.K.Y.; Writing–review and editing, A.K.Y.

Funding: This research received no external funding.

Acknowledgments: This work was financially supported by BK 21 PLUS, Creative Human Resource Development Program for IT Convergence (NRF-2015R1A4A1041584), Pusan National University, Busan, South Korea. We would like to thank KBSI, Busan for SEM, TEM, XRD, XPS and EDX analysis.

Conflicts of Interest: The authors declare no conflict of interest.

References

1. Liao, J.; Zou, P.; Su, S.; Nairan, A.; Wang, Y.; Wu, D.; Wong, C.-P.; Kang, F.; Yang, C. Hierarchical nickel nanowire@NiCo$_2$S$_4$ nanowhisker composite arrays with a test-tube-brush-like structure for high-performance supercapacitors. *J. Mater. Chem. A* **2018**, *6*, 15284–15293. [CrossRef]
2. Cai, D.; Huang, H.; Wang, D.; Liu, B.; Wang, L.; Liu, Y.; Li, Q. High performance supercapacitor electrode based on the unique ZnO@Co$_3$O$_4$ core/shell heterostructures on Nickel foam. *ACS Appl. Mater. Interfaces* **2014**, *6*, 15905–15912. [CrossRef] [PubMed]

3. Gobal, F.; Faraji, M. Preparation and electrochemical performances of nanoporous/cracked cobalt oxide layer for supercapacitors. *Appl. Phys. A* **2014**, *117*, 2087–20194. [CrossRef]

4. Kumar, Y.A.; Rao, S.S.; Punnose, D.; Tulasivarma, C.V.; Gopi, C.V.V.M.; Prabakar, K.; Kim, H.-J. Influence of solvents in the preparation of cobalt sulfide for supercapacitor. *R. Soc. Open Sci.* **2017**, *4*, 170427. [CrossRef] [PubMed]

5. Zhang, Y.; Ma, M.; Yang, J.; Su, H.; Huang, W.; Dong, X. Selective synthesis of hierarchical mesoporous spinal NiCo$_2$O$_4$ for high-performance supercapacitors. *Nanoscale* **2014**, *6*, 4303–4308. [CrossRef] [PubMed]

6. Chen, Z.; Wan, Z.; Yang, T.; Zhao, M.; Lv, X.; Wang, H.; Ren, X.; Mei, X. Preparation of nickel cobalt sulfide hollow nanocolloids with enhanced electrochemical property for supercapaictors applications. *Sci. Rep.* **2014**, *6*, 25151. [CrossRef] [PubMed]

7. Meng, X.; Sun, H.; Zhu, J.; Bi, H.; Han, Q.; Liu, X.; Wang, X. Graphene-based cobalt sulfide composite hydrogel with enhanced electrochemical properties for supercapcitors. *New J. Chem.* **2016**, *40*, 2843–2849. [CrossRef]

8. Kandalkar, S.G.; Gunjakar, J.L.; Lokhande, C.D. Chemical synthesis of cobalt oxide thin film electrode for supercapacitor application. *Appl. Surf. Sci.* **2008**, *254*, 5540–5544. [CrossRef]

9. Liu, S.; Mao, C.; Niu, Y.; Yi, F.; Hou, J.; Lu, S.; Jiang, J.; Xu, M.; Li, C. Facile synthesis of novel networked ultralong cobalt sulfide nanotubes and its application in supercapacitors. *ACS Appl. Mater. Interfaces* **2015**, *7*, 25568–25573. [CrossRef] [PubMed]

10. Yedluri, A.K.; Kim, H.-J. Wearable super-high specific performance supercapacitors using a honeycomb with folded silk-like composite of NiCo$_2$O$_4$ nanoplates decorated with NiMoO$_4$ honeycombs on nickel foam. *Dalton Trans.* **2018**, *47*, 15545–15554. [CrossRef] [PubMed]

11. Huang, K.-J.; Zhang, J.Z.; Shi, G.W.; Liu, Y.-M. One step hydrothermal synthesis of two-dimentional cobalt sulfide for high performance supercapacitors. *Mater. Lett.* **2014**, *131*, 45–48. [CrossRef]

12. Lin, J.-Y.; Chou, S.-W. Cathodic deposition of interlaced nanosheet-like cobalt sulfide films for high-performance supercapacitors. *RSC Adv.* **2013**, *3*, 2043–2048. [CrossRef]

13. Cheng, J.; Lu, Y.; Qiu, K.; Yan, H.; Xu, J.; Han, L.; Liu, X.; Luo, J.; Kim, J.K.; Luo, Y. Hierarchical core/shell NiCo$_2$O$_4$@NiCo$_2$O$_4$ Nanocactus arrays with dual-functionalities for high-performance supercapacitors and Li-ion batteries. *Sci. Rep.* **2015**, *5*, 12099. [CrossRef] [PubMed]

14. Yu, Z.; Tetard, L.; Zhai, L.; Thomas, J. Supercapacitor electrode materials: nanostructures from 0 to 3 dimensions. *Energy Environ. Sci.* **2015**, *8*, 702–730. [CrossRef]

15. Dong, W.; Wang, X.; Li, B.; Wang, L.; Chen, B.; Li, C.; Li, X.; Zhang, T.; Shi, Z. Hydrothermal synthesis and structure evolution of hierarchical cobalt sulfide nanostructures. *Dalton Trans.* **2011**, *40*, 243–248. [CrossRef] [PubMed]

16. Liu, R.; Ma, L.; Huang, S.; Mei, J.; Li, E.; Yuan, G. Large areal mass and high scalable and flexible cobalt Oxide/Graphene/Bacterial cellulose electrode for supercapacitors. *J. Phys. Chem. C* **2016**, *120*, 28480–28488. [CrossRef]

17. Tao, F.; Zhao, Y.-Q.; Zhang, G.-Q.; Li, H.-L. Electrochemical characterization on cobalt sulfide for electrochemical supercapacitors. *Electrochem. Commun.* **2007**, *9*, 1282–1287. [CrossRef]

18. Li, Z.; Han, J.; Fan, L.; Guo, R. Template-free synthesis of Ni$_7$S$_6$ hollow spheres with mesoporous shells for high performance supercapacitors. *CrystEngComm* **2015**, *17*, 1952–1958. [CrossRef]

19. Wang, X.; Zhao, S.-X.; Dong, L.; Lu, Q.-L.; Zhu, J.; Nan, C.-W. One-step synthesis of surface-enriched nickel cobalt sulfide nanoparticles on grapheme for high-performance supercapacitors. *Energy Storage Mater.* **2017**, *6*, 180–187. [CrossRef]

20. Kumar, R.; Kim, H.-J.; Park, S.; Srivastava, A.; Oh, I.-k. Graphene-wrapped and cobalt oxide-intercalated hybrid for extremely durable super-capacitor with ultrahigh energy and power densities. *CARBON* **2014**, *79*, 192–202. [CrossRef]

21. Patil, D.S.; Pawar, S.A.; Shin, J.C. Core-shell structure of Co$_3$O$_4$@CdS for high performance electrochemical supercapacitor. *Chemical Engineering Journal.* **2018**, *335*, 693–702. [CrossRef]

22. Yedluri, A.K.; Kim, H.-J. Wearable super-high specific performance supercapacitors using a honeycomb with folded silk-like composite of NiCo$_2$O$_4$ nanoplates decorated with NiMoO$_4$ honeycombs on nickel foam. *Dalton Trans.* **2018**, *47*, 15545–15554. [CrossRef] [PubMed]

23. Singh, A.K.; Sarkar, K.K.; Mandal, K.; Khan, G.G. High-performance supercapacitor electrode based on cobalt oxide-manganese dioxide-nickel oxide ternary 1D hybrid nanotubes. *ACS Appl. Mater. Interfaces* **2016**, *8*, 20786–20792. [CrossRef] [PubMed]

24. Yao, M.; Hu, Z.; Liu, Y.; Liu, P. Design synthesis of hierarchical $NiCo_2S_4@NiMoO_4$ core/shell nanospheres for high-performance supercapacitors. *New J. Chem.* **2015**, *39*, 8430. [CrossRef]

25. Jiang, J.; Shi, W.; Song, S.; Hao, Q.; Fan, W.; Xia, X.; Zhang, X.; Wang, Q.; Liu, C.; Yan, D. Solvothermal synthesis and electrochemical performance in super-capacitor of Co_3O_4/C flower-like nanostructure. *J. Power Sources* **2014**, *248*, 1281–1289. [CrossRef]

26. Selvam, S.; Balamuralitharan, B.; Jegatheeswaran, S.; Kim, M.Y.; Karthick, S.N.; Raj, J.A.; Boomi, P.; Sundrarajan, M.; Prabakar, K.; Kim, H.J. Electrolyte-imprinted grpahene oxide-chitosan chelate with copper crosslinked composite electrodes for intense cyclic-stable flexible supercapacitors. *J. Mater. Chem. A* **2017**, *5*, 1380–1386. [CrossRef]

27. Gu, Z.; Zhang, X. $NiCo_2O_4@MnMoO_4$ core-shell flowers for high performance supercapacitors. *J. Mater. Chem. A* **2016**, *4*, 8249. [CrossRef]

28. Bai, J.; Wang, K.; Feng, J.; Xiong, S. ZnO/CoO $ZnCo_2O_4$ Hierarchical Bipyramid Nanoframes: Morphology Control, Formation Mechanism, and Their Lithium Storage Properties. *ACS Appl. Mater. Interfaces* **2015**, *7*, 22848–22857. [CrossRef] [PubMed]

29. Yedluri, A.K.; Kim, H.-J. Wearable super-high specific performance supercapacitors using a honeycomb with folded silk-like composite of $NiCo_2O_4$ nanoplates decorated with $NiMoO_4$ honeycombs on nickel foam. *Dalton Trans.* **2018**, *47*, 15545–15554. [CrossRef] [PubMed]

30. Niu, H.; Zhou, D.; Yang, X.; Li, X.; Wang, Q.; Qu, F. Towards three-dimensional hierarchical ZnO nanofiber@Ni(OH)$_2$ nanoflake core-shell heterostructures for high-performance asymmetric supercapacitors. *J. Mater. Chem. A* **2015**, *3*, 18413. [CrossRef]

31. Du, J.; Zhou, G.; Zhang, H.; Cheng, C.; Ma, J.; Wei, W.; Chen, L.; Wang, T. Ultrthin porous $NiCo_2O_4$ nanosheet arrays on flexible carbon fabric for high-performance supercapacitors. *ACS Appl. Mater. Interfaces* **2013**, *5*, 7405–7409. [CrossRef] [PubMed]

32. Yuan, C.; Li, J.; Hou, L.; Zhang, X.; Shen, L.; Lou, X.W. Ultrathin mesoporous $NiCo_2O_4$ nanosheets supported on Ni foam as advanced electrodes for supercapcitors. *Adv. Funct. Mater.* **2012**, *22*, 4592−4597. [CrossRef]

33. Pujari, R.B.; Lokhande, A.C.; Kim, J.H.; Lokhande, C.D. Bath temperature controlled phase stability of hierarchical nanoflakes CoS_2 thin films for supercapacitor applications. *RSC Adv.* **2016**, *6*, 40593–40601. [CrossRef]

34. Xiao, X.; Han, B.; Chen, G.; Wang, L.; Wang, Y. Praparation and electrochemical performnaces of carbon sphere@ZnO core-shell nanocomposites for supercapacitor applications. *Sci. Rep.* **2017**, *7*, 40167. [CrossRef] [PubMed]

35. Zheng, Q.; Kvit, A.; Cai, Z.; Ma, Z.; Gong, Z. A free standing cellulose nnaofibril-reduced grapheme oxide-molybdenum oxynitride aerogel film electrode for all-solid-state supercapacitors with ultrahigh energy density. *J. Mater. Chem. A* **2017**, *5*, 12528–12541. [CrossRef]

36. Mahajan, H.; Bae, J.; Yun, K. Facile synthesis of ZnO-Au composites for high performance supercapacitors. *J. Alloys Compd.* **2018**, *758*, 131. [CrossRef]

37. Xu, J.; Liu, S.; Liu, Y. Co_3O_4/ZnO nanoheterostructure derived from core-shell ZIF-8@ZIF-67 for supercapacitors. *RSC Adv.* **2016**, *6*, 52137. [CrossRef]

38. Gai, Y.; Shang, Y.; Gong, L.; Su, L.; Hao, L.; Dong, F.; Li, J. A self-template synthesis of porous $ZnCo_2O_4$ microspheres for high-performance quasi-solid-state asymmetric supercapacitors. *RSC Adv.* **2017**, *7*, 1038–1044. [CrossRef]

39. Xu, L.; Zhao, Y.; Lian, J.; Xu, Y.; Bao, J.; Qiu, J.; Xu, L.; Xu, H.; Hua, M.; Li, H. Morphology controlled preparation of $ZnCo_2O_4$ nanostructures for asymmetric supercapacitor with ultrahigh energy density. *Energy* **2017**, *123*, 296–304. [CrossRef]

40. Mohapatra, D.; Parida, S.; Badrayyana, S.; Singh, B.K. High performance flexible asymmetric CNO-ZnO//ZnO supercapacitor with an operating voltage of 1.8V in aqueous medium. *Appl. Mater. Today* **2017**, *7*, 212–221. [CrossRef]

41. Xu, J.; Gao, L.; Cao, J.; Wang, W.; Chen, Z. Preparation and electrochemical capacitance of cobalt oxide (Co_3O_4) nanotubes as supercapacitor material. *Electrochim. Acta* **2010**, *56*, 732–736. [CrossRef]

42. Pant, B.; Park, M.; Ojha, G.P.; Park, J.; Kuk, Y.S.; Lee, E.J.; Kim, H.Y.; Park, S.J. carbon nanofibers wrapped with zinc oxide nanoflakes as promising electrode for supercapacitors. *J. Coll. Interface Sci.* **2018**, *522*, 40–47. [CrossRef] [PubMed]

43. Cui, L.; Li, J.; Zhang, X.G. Preparation and properties of Co_3O_4 nanorods as supercapacitor material. *J. Appl. Electrochem.* **2009**, *39*, 1871. [CrossRef]

44. Wang, Q.; Zhu, L.; Sun, L.; Liub, Y.; Jiao, L. Facile synthesis of hierarchical porous $ZnCo_2O_4$ microspheres for high-performance supercapacitors. *J. Mater. Chem. A* **2015**, *3*, 982. [CrossRef]

45. Li, W.; Yang, F.; Hu, Z.; Liu, Y. Template synthesis of $C@NiCo_2O_4$ hollow microsphere as electrode material for supercapacitor. *J. Alloys Compd.* **2018**, *749*, 305–312. [CrossRef]

46. Yang, M.H.; Choi, B.G. Preparation of Three-Dimensional Co_3O_4/graphene Composite for High-Performance Supercapacitors. *Chem. Eng. Commun.* **2017**, *204*, 723–728. [CrossRef]

Article

Fabrication of Hierarchical NiMoO$_4$/NiMoO$_4$ Nanoflowers on Highly Conductive Flexible Nickel Foam Substrate as a Capacitive Electrode Material for Supercapacitors with Enhanced Electrochemical Performance

Anil Kumar Yedluri [ID]**, Tarugu Anitha and Hee-Je Kim ***

School of Electrical Engineering, Pusan National University, Busandaehak-ro 63beon-gil, Geumjeong-gu, Busan 46241, Korea; yedluri.anil@pusan.ac.kr (A.K.Y.); anitha.tarugu@gmail.com (T.A.)
*   Correspondence: heeje@pusan.ac.kr; Tel.: +82-10-2295-0613; Fax: +82-51-513-0212

Received: 26 January 2019; Accepted: 19 March 2019; Published: 24 March 2019

Abstract: Hierarchical NiMoO$_4$/NiMoO$_4$ nanoflowers were fabricated on highly conductive flexible nickel foam (NF) substrates using a facile hydrothermal method to achieve rapid charge-discharge ability, high energy density, long cycling lifespan, and higher flexibility for high-performance supercapacitor electrode materials. The synthesized composite electrode material, NF/NiMoO$_4$/NiMoO$_4$ with a nanoball-like NF/NiMoO$_4$ structure on a NiMoO$_4$ surface over a NF substrate, formed a three-dimensional interconnected porous network for high-performance electrodes. The novel NF/NiMoO$_4$/NiMoO$_4$ nanoflowers not only enhanced the large surface area and increased the electrochemical activity, but also provided an enhanced rapid ion diffusion path and reduced the charge transfer resistance of the entire electrode effectively. The NF/NiMoO$_4$/NiMoO$_4$ composite exhibited significantly improved supercapacitor performance in terms of a sustained cycling life, high specific capacitance, rapid charge-discharge capability, high energy density, and good rate capability. Electrochemical analysis of the NF/NiMoO$_4$/NiMoO$_4$ nanoflowers fabricated on the NF substrate revealed ultra-high electrochemical performance with a high specific capacitance of 2121 F g^{-1} at 12 mA g^{-1} in a 3 M KOH electrolyte and 98.7% capacitance retention after 3000 cycles at 14 mA g^{-1}. This performance was superior to the NF/NiMoO$_4$ nanoball electrode (1672 F g^{-1} at 12 mA g^{-1} and capacitance retention 93.4% cycles). Most importantly, the SC (NF/NiMoO$_4$/NiMoO$_4$) device displayed a maximum energy density of 47.13 W h kg^{-1}, which was significantly higher than that of NF/NiMoO$_4$ (37.1 W h kg^{-1}). Overall, the NF/NiMoO$_4$/NiMoO$_4$ composite is a suitable material for supercapacitor applications.

Keywords: supercapacitor; hierarchical nanostructures; NiMoO$_4$/NiMoO$_4$; NiMoO$_4$; flexible supercapacitor electrode; excellent cycling stability

1. Introduction

In recent years, novel energy sources have become a major challenge for global sustainable development in the 21st century due to the exhaustion of fossil fuels and environmental pollution [1–4]. The concerns regarding global warming caused by the use of non-renewable fossil fuels have increased considerably [5]. Over the past decade, serious efforts have been made to exploit renewable energy, such as biofuels, wind power, and tidal sources, with the intention of partially replacing fossil fuels [6–8]. In this regard, considering the various energy storage devices available, the most promising and fast emerging storage device is the supercapacitor (SC) owing to its high energy and power density, rapid charge-discharge capability, high flexibility, and sustained cycle life [9,10]. SCs can provide a clean and efficient emerging storage technology to help solve the global and environmental energy crises [11]. Generally, energy storage in SCs takes place via two methods: electrochemical double layer capacitors (EDLCs) and pseudocapacitors (PCs or redox capacitors) [12,13]. In EDLCs, the capacitance is due to ion adsorption and desorption at the surface of the carbon electrode and redox electrolyte, whereas in PCs, the energy is stored in reversible faradaic redox reactions taking place at the electrolyte-electrode interface [14]. PCs generally have a very high energy density and specific capacitance compared to EDLCs, owing to their excellent conductivity and electrochemical properties [15,16].

Of the different types of redox-active materials, ternary transition metal oxides (TMOs) are used widely as PC electrode materials because of their various oxidation states for faradaic redox reactions and high theoretical capacitance. Currently, ternary transition metal oxides, such as $NiCo_2O_4$, $CoMoO_4$, $MnMoO_4$, $ZnCo_2O_4$, $MnCo_2O_4$, $CuCo_2O_4$, and $NiMoO_4$, have been studied extensively as new types of energy storage materials owing to their excellent electrochemical performance and better conductivity [17–23]. Recently, few studies have examined the design of multicomponent nanostructured materials for SC applications that show excellent energy storage behavior. Lou et al. prepared $NiCo_2O_4$ structures exhibiting a nanorod and nanosheet morphology using a facile solution method, which delivered a specific capacitance (Cs) of 905 and 887.7 F g^{-1} at 2 A g^{-1} [24]. Manickam et al. used a solution combustion synthesis approach to develop α-$NiMoO_4$ nanoparticles, exhibiting a high Cs of 1517 F g^{-1} at 1.2 A g^{-1} [25]. Furthermore, Zhang et al. reported a nanowire array of $NiCo_2O_4$ using a simple hydrothermal technique for SC applications, showing a super high Cs of 1283 F g^{-1} at 1 A g^{-1} [26]. On the other hand, combining two types of TMOs to form a unique nanostructure is an efficient way to improve the EC performance further. For example, Mai et al. synthesized $MnMoO_4$-$CoMoO_4$ nanowire structures using simple refluxing technology and achieved a high Cs of 181 F g^{-1} at 1 A g^{-1} [27]. Wang et al. used a solution method to synthesize $NiCo_2O_4$@$MnMoO_4$ nanocolumn arrays over a Ni foam substrate, exhibiting a high Cs of 1705.3 F g^{-1} at 5 mA cm^{-2} [28]. Recently, Cheng et al. reported hierarchical core/shell $NiCo_2O_4$@$NiCo_2O_4$ nanocactus arrays for SCs applications, showing a very high Cs of 1264 F g^{-1} at 2 A g^{-1} [29].

Among the various TMOs available, nickel molybdate ($NiMoO_4$) has attracted enormous attention owing to its outstanding electrochemical conductivity, high theoretical capacity, cost-effectivity, rapid redox activity, low cost, earth abundance, and non-toxicity [30–35]. Until now, very few studies have evaluated $NiMoO_4$/$NiMoO_4$ as an electrode material for SC applications. The main challenge for potential real-time applications of SCs is to achieve high energy density, higher specific capacitance, sustained cycle life, rate capability, high flexibility, and fast charge-discharge capability [36]. Therefore, the development of a simple technique to fabricate nanostructured $NiMoO_4$/$NiMoO_4$ composite electrode materials with unique morphologies and high energy storage performance will be of great significance.

Based on the above ideas, an aligned compact of a NF/$NiMoO_4$ electrode and a NF/$NiMoO_4$/$NiMoO_4$ composite on a nickel foam substrate were synthesized using a simple hydrothermal approach to produce highly efficient electrode materials. The use of nickel foam as a current collector increases the connection between $NiMoO_4$/$NiMoO_4$ nanostructures with the collectors. In addition, nickel foam (NF) also improves the conductivity, flexibility, and electron/ion transportation. This hierarchical NF/$NiMoO_4$/$NiMoO_4$ composite allowed the migration of electrolyte ions into the inner regions of the electrode, which could reduce the resistance of the electrolyte. The electrochemical behaviors of the hierarchical structured NF/$NiMoO_4$/$NiMoO_4$ composite electrode were investigated in a three-electrode system using a 3 M KOH aqueous electrolyte by cyclic voltammetry (CV), galvanostatic charge-discharge (GCD), and electrochemical impedance spectroscopy (EIS). The NF/$NiMoO_4$/$NiMoO_4$ composite, as the active electrode material, exhibited a high specific capacitance of 2121 F g^{-1} at a current density of 12 mA g^{-1}, and excellent cycling stability (98.7% capacitance retention after 3000 cycles). The high performance observed was attributed to the synergism between NF/$NiMoO_4$ and NF/$NiMoO_4$. To the best of the authors' knowledge, the outstanding performance of the NF/NiMoO$_4$/NiMoO$_4$ composite was superior to that of the NF/NiMoO$_4$ electrode material, making it a promising candidate as a SCs material.

2. Experimental Methods

2.1. Chemicals and Materials

Sodium molybdate dihydrate [Na$_2$MoO$_4$.2H$_2$O], nickel nitrate hexahydrate [Ni(NO$_3$)$_2$.6H$_2$O], thiourea [CH$_4$N$_2$S], ammonium fluoride [NH$_4$F], potassium hydroxide [KOH], and Ni foam were obtained from Sigma-Aldrich Co., South Korea. Deionized (DI) water was used throughout the study. All chemicals used for the synthesis of the NF/NiMoO$_4$ nanoball structures and NF/NiMoO$_4$/NiMoO$_4$ nanoflower-like structures were of analytical grade and used as received.

2.2. Preparation of the NF/NiMoO$_4$ Nanoball Electrode on a Ni Foam Substrate

The composite NF/NiMoO$_4$/NiMoO$_4$ nanoflower electrode material was grown on Ni foam using a facile hydrothermal reaction. Before the synthesis process, the Ni foam (1.5 $\times$ 2 cm^2) was treated with 3 M HCl for 40 min and were then cleaned with deionized (DI) water and ethanol for 30 min each. The host precursor materials, such as 0.03 mol of [Ni(NO$_3$)$_2$.6H$_2$O] and 0.06 mol of [Na$_2$MoO$_4$.2H$_2$O], were dissolved in 100 mL of distilled water and were stirred for 1 h. Subsequently, 0.4 mol thiourea [CH$_4$N$_2$S] and 0.04 mol ammonium fluoride [NH$_4$F] were added to the above mixture whilst stirring. The mixture was continuously stirred for 1 h until complete dissolution. The above solution and a piece of Ni foam were transferred to a Teflon-lined stainless steel autoclave. The hydrothermal reaction system was conducted at 140 °C for 12 h. After cooling to room temperature, the NF/NiMoO$_4$-loaded Ni foam electrode was cleaned with DI water and ethanol, and dried overnight in a vacuum oven at 100 °C. The collected NF/NiMoO$_4$ foams were then annealed at 200 °C for 5 h to obtain the final composite material.

2.3. Preparation of the NF/NiMoO$_4$/NiMoO$_4$ Nanoflower Composite Electrode on a Ni Foam Substrate

A secondary hydrothermal reaction was adopted; NF/NiMoO$_4$ was used as the scaffold on the nickel foam for further growth of the NF/NiMoO$_4$/NiMoO$_4$ structure using a similar process. Typically, 0.03 mol of [Ni(NO$_3$)$_2$.6H$_2$O] and 0.06 mol of [Na$_2$MoO$_4$.2H$_2$O] were dissolved in 100 mL of DI water with magnetic stirring for 20 min. The above solution was then transferred to a Teflon-lined stainless steel autoclave and the NF/NiMoO$_4$-deposited nickel foams were placed into the reaction mixture and maintained at 120 °C for 8 h. The autoclave was cooled naturally to room temperature and the as-obtained NF/NiMoO$_4$/NiMoO$_4$ composite material was washed with DI water and ethanol, and finally dried overnight at 80 °C. Finally, the samples were annealed in air at 250 °C for 5 h. The mass loading of the NF/NiMoO$_4$ and NF/NiMoO$_4$/NiMoO$_4$ nanostructures were around 7.0 and 10 mg cm^{-2}, respectively.

2.4. Characterizations of Nanoballs and Nanoflowers

All samples were examined by field emission scanning electron microscopy (FE-SEM, S-4800, Hitachi) at the Busan KBSI, high-resolution transmission electron microscopy (HRTEM, Jem 2011, Jeol corp. Tokyo, Japan), high resolution X-ray diffraction (XRD, Bruker D8 Advance, Tokyo, Japan) using Cu Kα operated at 40 kV and 30 mA, and X-ray photoelectron spectroscopy (XPS, VG scientific ESCALAB 250 system, Hitachi manufacturer, Tokyo, Japan).

2.5. Electrochemical Measurements

Electrochemical characterization of the as-prepared (NF/NiMoO$_4$ and NF/NiMoO$_4$/NiMoO$_4$) electrodes, such as CV, GCD measurements, and EIS, was carried out in a three-electrode configuration using a Biologic-SP150 workstation. A three-electrode glass cell setup was used in the experiment. The prepared samples, platinum wire, and saturated Ag/AgCl were used as the working, counter, and reference electrodes, respectively, with a 3.0 M KOH aqueous solution as the supporting electrolyte. A charge-discharge study was carried out galvanostatically at 12–18 mA g^{-1} from 0 to 0.4 V vs. Ag/AgCl and CV was recorded in the potential window of -0.2 to 0.5 V vs. Ag/AgCl by varying the scan rate from 1 to 30 mV s^{-1}. The specific capacitance, energy density and power density were calculated from the GCD curves at different current densities according to the following Equations (1)–(3) [37]:

$$C_s = \frac{I \times \Delta t}{m \times \Delta V} \tag{1}$$

$$E = \frac{C_s \times (\Delta V^2)}{2} \tag{2}$$

$$P = \frac{E}{\Delta t} \tag{3}$$

where I (A) represents current density; the potential range is depicted as ΔV. m (g) denotes the active material mass and discharge time is depicted as Δt (s) [38].

3. Results and Discussion

Preparation, Morphology, and Characterization of NF/NiMoO$_4$ and NF/NiMoO$_4$/NiMoO$_4$

The surface morphologies of the synthesized NF/NiMoO$_4$ and NF/NiMoO$_4$/NiMoO$_4$ were examined by FESEM. Figure 1a–f presents the FESEM images (lower and higher magnifications) of NF/NiMoO$_4$ and NF/NiMoO$_4$/NiMoO$_4$ synthesized on a nickel foam substrate using a facile hydrothermal method. Figure 1a–c shows SEM images of NF/NiMoO$_4$, revealing the formation of ball-like structures with rough NiMoO$_4$ nanostructures on the NF surface. The image revealed a large number of voids among the balls that were not distributed uniformly over the NiMoO$_4$ nanostructure on NF (Figure 1b), which is not a beneficial surface for electrochemical reactions (Figure 1c). In contrast, the FE-SEM image of NF/NiMoO$_4$/NiMoO$_4$ nanoflowers showed a three-dimensional interconnected, porous, and roughened surface structure without major voids, where ultrathin nanoballs of NF/NiMoO$_4$ were grown and were distributed uniformly over the surface of the NiMoO$_4$ nanostructure (Figure 1d,f). The three-dimensional nanostructures interconnected with each other to form a highly porous structure that was expected to enhance the surface area, increase the electrochemical conductivity of the working materials and facilitate rapid transportation between the active substance and electrolyte ion in supercapacitor electrodes.

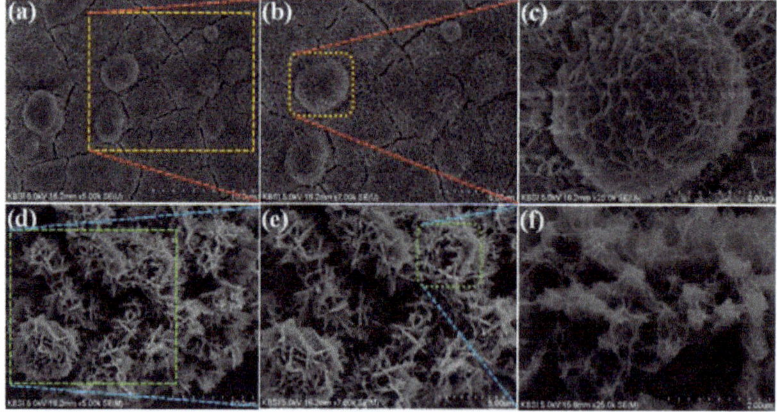

Figure 1. Morphological characterization of Nickel foam NF/NiMoO$_4$ and NF/NiMoO$_4$/NiMoO$_4$. FE-SEM images of the prepared (**a–c**) NF/NiMoO$_4$ and (**d–f**) of NF/NiMoO$_4$/NiMoO$_4$ and their corresponding high magnification images.

Figure 2a,b shows different magnification HR-TEM and TEM images of NF/NiMoO$_4$, displaying similar structures, which are in good agreement with FE-SEM analysis. Figure 1b presents a TEM image with dark strips with a crumpled silk-like structure. The crystalline lattice fringe had a spacing of 0.31 nm and was assigned to the (220) plane of NiMoO$_4$. EDX analysis of NF/NiMoO$_4$ (Figure 2c) revealed the presence of Ni, Mo, and O elements without impurities. The atomic percentage of elements of Mo, Ni, O and C are obtained to be 13.34%, 44.30%, 37.51% and 4.82%, respectively. On the other hand, TEM images of the NF/NiMoO$_4$ nanoballs in the NF/NiMoO$_4$/NiMoO$_4$ electrode revealed a lattice fringe spacing of 0.352 nm, which was assigned to the (111) plane of NiMoO$_4$, as shown in Figure 2d,e. The NF/NiMoO$_4$/NiMoO$_4$ nanoflowers were comprised mainly of Ni, Mo, and O (Figure 2f) arising from the nickel foam, which confirmed the successful fabrication of NF/NiMoO$_4$/NiMoO$_4$. The atomic percentage of elements of Mo, Ni, O and C were obtained as 29.44%, 34.27%, 37.48% and 3.78%, respectively.

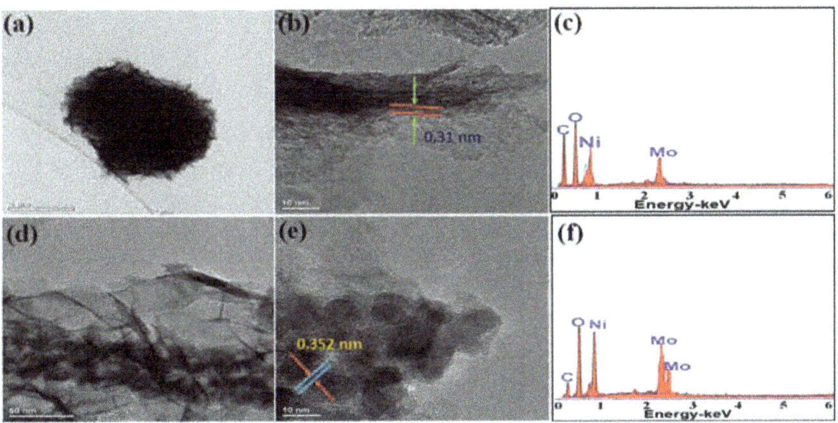

Figure 2. (**a,b**) HR-TEM and TEM images and (**c**) EDX pattern of NF/NiMoO$_4$ on nickel foam. (**d,e**) TEM images and (**f**) Energy-dispersive X-ray spectroscopy (EDX) pattern of NF/NiMoO$_4$/ NiMoO$_4$ nanostructure on nickel foam.

The phase purity and crystal nature of the NF/NiMoO$_4$ and NF/NiMoO$_4$/NiMoO$_4$ electrodes were analyzed further by XRD. Both samples exhibited XRD peaks at 2θ angle of 14.3°, 25.3°, 28.8°, 32.6° and 43.9° 2θ (Figure 3), which were assigned to (1 1 0), (1 1 2), (2 2 0), (0 2 2), and (3 3 0) diffraction planes of the monoclinic phase of NiMoO$_4$ (JCPDS NO. 33-0948) [39]. Some other XRD peaks from the nickel foam substrate were observed. Overall, the XRD data concurred with the TEM observations.

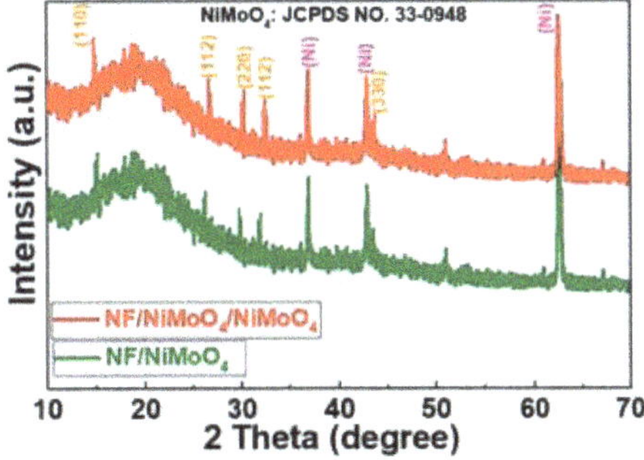

Figure 3. XRD pattern of the fabricated NF/NiMoO$_4$ electrode and NF/NiMoO$_4$/NiMoO$_4$ composite electrode grown on NF.

The chemical compositions and oxidation states of the elements of NF/NiMoO$_4$/NiMoO$_4$ were analyzed further by XPS analysis and the results are shown in Figure 4. The survey spectrum of NF/NiMoO$_4$/NiMoO$_4$ revealed the presence of Mo 3d, Ni 2p, and O 1s. The XPS spectra of Mo 3d (Figure 4b) showed two major peaks centered at binding energies of 232 and 235 eV with a spin energy separation of 3 eV, corresponding to Mo 3d$_{5/2}$ and Mo 3d$_{3/2}$, respectively [40]. De-convolution of the Ni 2p spectrum (Figure 4b) revealed two prominent Ni 2P$_{3/2}$ and Ni 2P$_{1/2}$ spin-orbit peaks at 855.3 and 872.6 eV respectively; each were fitted with spin-orbit doublet characteristic of the Ni^{3+} and Ni^{2+} binding states [41]. Figure 4c presents the de-convolution of the O 1s spectra showing the oxygen contributions at 530.9 eV and 532 eV, which were assigned to s O$_1$ and O$_2$ associated with typical metal oxygen bonds [42,43].

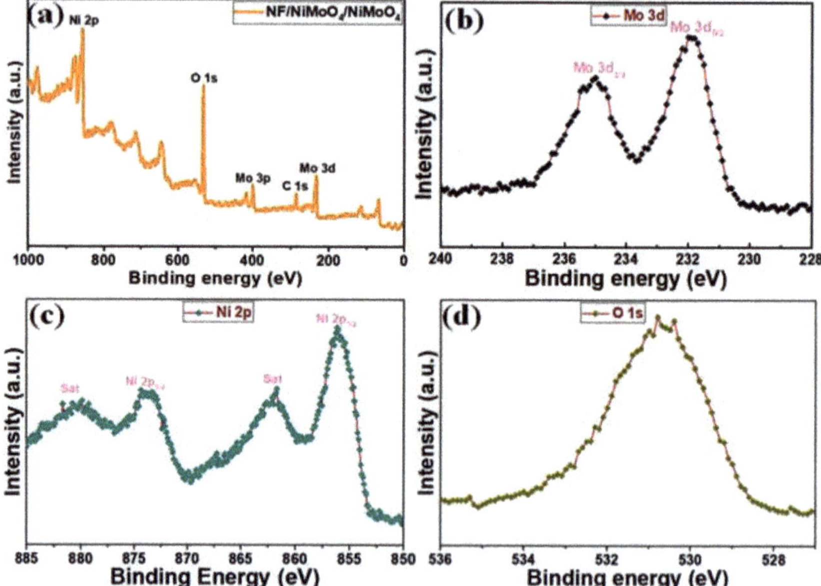

Figure 4. XPS survey spectrum (**a**), Mo 3d (**b**), Ni 2p (**c**) and O 1s (**d**) of the NF/NiMoO$_4$/NiMoO$_4$ composite material.

The electrochemical performance of the as-obtained samples of NF/NiMoO$_4$ and NF/NiMoO$_4$/NiMoO$_4$ were examined by CV and GCD in a three-electrode cell configuration with a 3 M KOH aqueous electrolyte. Figure 5a presents the typical CV curves of the NF/NiMoO$_4$ and NF/NiMoO$_4$/NiMoO$_4$ electrodes at a scan rate of 5 mV s^{-1} in the potential window of -0.2 to 0.5 V (vs. saturated calomel electrode (SCE)). Figure 5b,c presents the typical CV curves of the NF/NiMoO$_4$/NiMoO$_4$ and NF/NiMoO$_4$ material on nickel foam substrates at various scan rates from 1 to 30 mV s^{-1} over the potential range of -0.2 to 0.5 V. All CV curves revealed a well-defined pair of redox peaks, which could be attributed to a Faradic redox reaction. The CV integrated area, oxidation, and reduction currents of the NF/NiMoO$_4$/NiMoO$_4$ electrode were much higher than those of the NF/NiMoO$_4$ electrode, highlighting the significant energy storage performance of the NF/NiMoO$_4$/NiMoO$_4$. Moreover, as the scan rate was increased from 1 to 30 mV s^{-1} for NF/NiMoO$_4$/NiMoO$_4$, the cathodic and anodic peak positions shifted to a lower and higher potential, showing superior electrochemical performance and higher specific capacitance than the individual components.

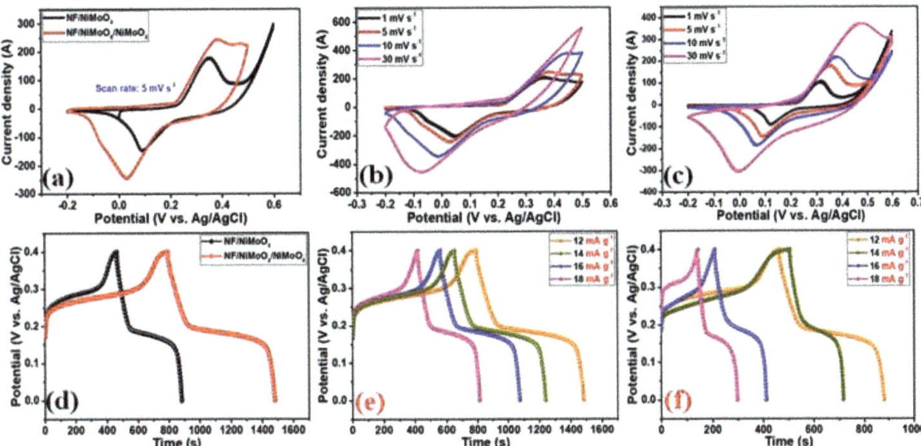

Figure 5. (a) Comparison of the CV curves of the NF/NiMoO$_4$ and NF/NiMoO$_4$/NiMoO$_4$ at the same scan rate (5 mV s^{-1}). (**b,c**) CV curves of the NF/NiMoO$_4$/NiMoO$_4$ electrode and NF/NiMoO$_4$ electrode at various scan rates. (**d**) GCD curves of the NF/NiMoO$_4$ and NF/NiMoO$_4$/NiMoO$_4$ electrodes at a current density of 12 mA g^{-1}. (**e,f**) Galvanostatic charge-discharge curves of the NF/NiMoO$_4$/NiMoO$_4$ and NF/NiMoO$_4$ electrodes at different current densities.

GCD was conducted to examine the high charge-storage behavior of NF/NiMoO$_4$/NiMoO$_4$ electrode compared to other electrode of NF/NiMoO$_4$. Figure 5d compares typical GCD plots of the NF/NiMoO$_4$ and NF/NiMoO$_4$/NiMoO$_4$ samples at 12 mA g^{-1} in the potential window of 0 to 0.4 V; significant differences were observed between the electrode materials. The NF/NiMoO$_4$/NiMoO$_4$ composite displayed a larger peak area than NF/NiMoO$_4$ electrode, highlighting its superior performance, which is in agreement with the CV result. Figure 5e,f presents GCD plots of NF/NiMoO$_4$/NiMoO$_4$ and NF/NiMoO$_4$ at different current densities from 12 to 18 mA g^{-1} in the potential window from 0 to 0.4 V. The GCD results were in good agreement with the CV data.

The gravimetric specific capacitance of the NF/NiMoO$_4$ and NF/NiMoO$_4$/NiMoO$_4$ electrode materials were determined from the discharge curves in a three-electrode configuration using Equation (1), as shown in the Figure 6a. The specific capacitance of the NF/NiMoO$_4$/NiMoO$_4$ and NF/NiMoO$_4$ electrodes were 2121, 2038, 1942, and 1806 F g^{-1} and 1672, 1358, 1260, and 1092 F g^{-1}, respectively, at a current density of 12, 14, 16 and 18 mA g^{-1}. The specific capacitance of the NF/NiMoO$_4$/NiMoO$_4$ electrode was significantly higher than that of the NF/NiMoO$_4$ electrode. This was attributed to the larger effective surface area, which provided improved accessibility to the electrolyte, enabled the effective transport of ions and reduced the charge transfer resistance. Furthermore, the as-obtained specific capacitance results of the NF/NiMoO$_4$/NiMoO$_4$ composite electrode in the present study is superior to those reported for NiMoO$_4$-based materials, such as hierarchical NiCo$_2$S$_4$@NiMoO$_4$ core/shell nanospheres (1714 F g^{-1} at 1 A g^{-1}) [44], Co$_3$O$_4$@NiMoO$_4$ nanosheets (1526 F g^{-1} at 3 mA cm^{-2}) [45], CoMoO$_4$-NiMoO$_4$.xH$_2$O bundles (1039 F g^{-1} at 2.5 mA cm^{-2}) [46], Co$_3$O$_4$@NiMoO$_4$ nanowires (1230 F g^{-1} at 10 mA cm^{-2}) [47], NiMoO$_4$@CoMoO$_4$ hierarchical nanospheres (1601.6 F g^{-1} at 2 A g^{-1}) [48], and core-shell NiCo$_2$O$_4$@NiMoO$_4$ nanowires (1325.9 F g^{-1} 2 mA cm^{-2}) [49]. To further investigate the long-practical applicability of the NF/NiMoO$_4$ and NF/NiMoO$_4$/NiMoO$_4$ electrodes for supercapacitor applications, a cycling stability test was carried out in a 3 M KOH electrolyte at 14 mA g^{-1} for 3000 cycles, as shown in Figure 6b. After 3000 cycles, the NF/NiMoO$_4$/NiMoO$_4$ electrode maintained approximately 98.7% of its initial capacitance, whereas the NF/NiMoO$_4$ electrode exhibited 93.4% capacitance retention. These results show that the NF/NiMoO$_4$/NiMoO$_4$ composite has outstanding electrochemical stability.

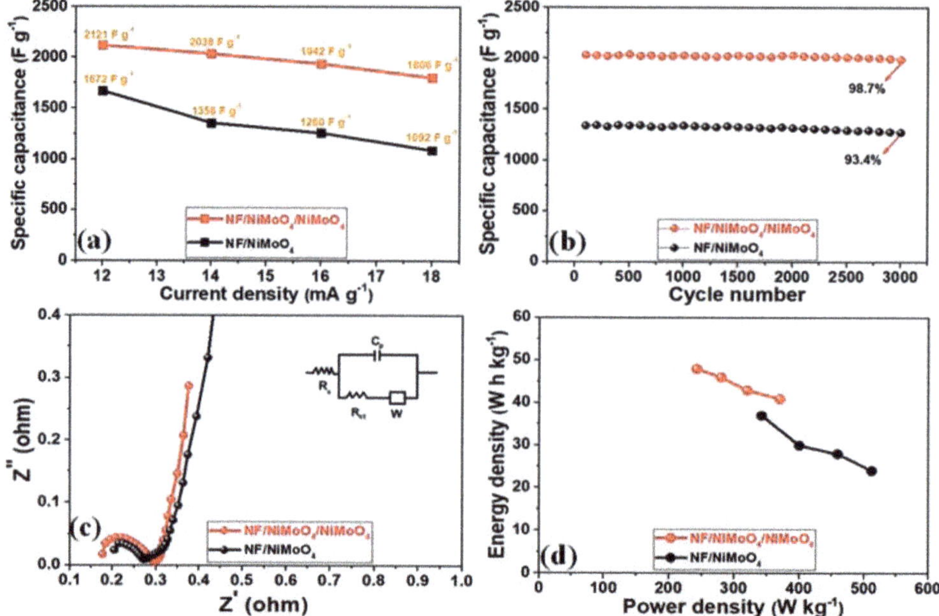

Figure 6. (**a**) Calculated specific capacitance at different current densities, (**b**) Long-term cycling stability, (**c**) Nyquist plots and (**d**) Ragone plot of the NF/NiMoO$_4$ and NF/NiMoO$_4$/NiMoO$_4$ electrode materials.

Figure 6c presents Nyquist plots of the NF/NiMoO$_4$/NiMoO$_4$ and NF/NiMoO$_4$ electrodes analyzed from the equivalent circuit from 0.01 Hz to 100 kHz. EIS data contains four categories: solution resistance R$_s$, charge transfer resistance R$_{ct}$, a pseudocapacitive element C$_p$, and Warburg element W. The R$_s$ can be estimated that the axis intercepts of the high frequency region in both active electrodes were almost similar. The R$_{ct}$, which distorted to be much smaller in the diameter of the semicircle arc, revealed that the NF/NiMoO$_4$/NiMoO$_4$ electrode consisted of lower interfacial charge-discharge impedance. In addition, the NF/NiMoO$_4$/NiMoO$_4$ electrode delivered a more vertical line, recommending the W which is not a determining factor and this type of sample can store charge more effectively. As expected, the R$_{ct}$ of NF/NiMoO$_4$/NiMoO$_4$ (0.05 Ω) was lower than that of NF/NiMoO$_4$ (0.09 Ω), which further confirmed the excellent conductivity of the NF/NiMoO$_4$/NiMoO$_4$ composite electrode. Figure 6d shows a Ragone plot (energy density vs. power density) of the as-obtained samples NF/NiMoO$_4$ and NF/NiMoO$_4$/NiMoO$_4$. The energy density and power density were calculated using Equations (2) and (3), respectively. The NF/NiMoO$_4$/NiMoO$_4$ device exhibited a very high energy density of 47.13 W h kg^{-1}, which was much higher than that of NF/NiMoO$_4$ (37.1 W h kg^{-1}), revealing excellent energy storage performance.

4. Conclusions

NF/NiMoO$_4$ and NF/NiMoO$_4$/NiMoO$_4$ nanostructures were prepared on a nickel foam substrate using a facile hydrothermal approach and evaluated as electrode materials for supercapacitor applications. The morphological and structural properties of the hierarchical NF/NiMoO$_4$/NiMoO$_4$ electrode material provides more active sites and large electrochemical surface area to enhance accessibility to the electroactive sites that improved the energy storing and transportation of ions and electrons compared to the NF/NiMoO$_4$ electrode. The electrochemical studies measured in the 3 M KOH electrolyte showed that this hierarchical NF/NiMoO$_4$/NiMoO$_4$ nanocomposite delivered superior electrochemical behavior with a maximum specific capacitance of 2121 F g^{-1} at 12 mA g^{-1} than the NF/NiMoO$_4$ electrode (1672 F g^{-1} at 12 mA g^{-1}), as well as outstanding cycling stability (98.7% and 93.4% capacitance retention after 3000 cycles at 14 mA g^{-1}). The NF/NiMoO$_4$/NiMoO$_4$ device possessed a higher energy density (47.13 W h kg^{-1}) than NF/NiMoO$_4$ (37.1 W h kg^{-1}). These features make the NF/NiMoO$_4$/NiMoO$_4$ composite electrode a potential candidate for next generation flexible energy storage systems.

Author Contributions: Conceptualization, A.K.Y.; formal analysis, T.A. and H.J.K.; investigation, A.K.Y. and H.J.K.; methodology, A.K.Y.; supervision, A.K.Y.; writing–original draft, A.K.Y.; writing–review and editing, A.K.Y.

Funding: This research received no external funding.

Acknowledgments: This work was financially supported by BK 21 PLUS, Creative Human Resource Development Program for IT Convergence (NRF-2015R1A4A1041584), Pusan National University, Busan, South Korea. We would like to thank KBSI, Busan for SEM, TEM, XRD, XPS and EDX analysis.

Conflicts of Interest: The authors declare no conflict of interest.

References

1. Li, Z.C.; Xin, Y.Q.; Jia, H.L.; Wang, Z.; Sun, J.; Zhou, Q. Rational design of coaxial MWCNT-COOH@NiCo$_2$S$_4$ hybrid for supercapacitors. *J. Mater. Sci.* **2017**, *52*, 9661–9672. [CrossRef]
2. Qin, G.; Zhang, H.Y.; Liao, H.Y.; Li, Z.H.; Tian, J.Y.; Lin, Y.X. Novel graphene nanosheet-wrapped polyaniline rectangular-like nanotubes for flexible all-solid-state supercapacitors. *J. Mater. Sci.* **2017**, *52*, 10981–10992. [CrossRef]
3. Zhao, J.; Lai, H.; Lyu, Z.Y.; Jiang, Y.F.; Xie, K.; Wang, X.Z. Hydrophilic hierarchical nitrogen-doped carbon nanocages for ultrahigh supercapacitive performance. *Adv. Mater.* **2015**, *27*, 3541–3545. [CrossRef]
4. Lee, J.H.; Kim, H.K.; Baek, E.; Pecht, M.; Lee, S.H.; Lee, Y.H. Improved performance of cylindrical hybrid supercapacitor using activated carbon/niobium doped hydrogen titanate. *J. Power Sources* **2016**, *301*, 348–354. [CrossRef]
5. Fajardy, M.; Dowell, N.M. Can BECCS deliver sustainable and resource efficient negative emissions? *Energy Environ. Sci.* **2017**, *10*, 1389–1426. [CrossRef]
6. Fares, R.L.; Webber, M.E. The impacts of storing solar energy in the home to reduce reliance on the utility. *Nat. Energy* **2017**, *2*, 17001. [CrossRef]
7. Segura, E.; Morales, R.; Somolinos, J.A.; López, A. Techno-economics challenges of tidal energy conversion system: Current status and trends. *Renew. Sustain. Energy Rev.* **2017**, *77*, 536–550. [CrossRef]
8. Wiser, R.; Jenni, K.; Seel, J.; Baker, E.; Hand, M.; Lantz, E.; Smith, A. Expert elicitations survey on future wind energy costs. *Nat. Energy* **2016**, *1*, 16135. [CrossRef]
9. Yedluri, A.K.; Kim, H.-J. Wearable super-high specific performance supercapacitors using a honeycomb with folded silk-like composite of NiCo$_2$O$_4$ nanoplates decorated with NiMoO$_4$ honeycombs on nickel foam. *Dalton Trans.* **2018**, *47*, 15545–15554. [CrossRef]
10. Yedluri, A.K.; Kim, H.-J. Preparation and electrochemical performance of NiCo$_2$O$_4$@NiCo$_2$O$_4$ composite nanoplates for high performance supercapacitor applications. *New J. Chem.* **2018**, *42*, 19971–19978.
11. Miller, J.R.; Simon, P. Electrochemical capacitors for energy management. *Science* **2008**, *321*, 651–652. [CrossRef]
12. Bai, M.H.; Bian, L.J.; Song, Y.; Liu, X.X. Electrochemical code position of vanadium oxide and polypyrrole for high-performance supercapacitor with high working voltage. *ACS Appl. Mater. Interface* **2014**, *6*, 12656–12664. [CrossRef] [PubMed]
13. Kumar, Y.A.; Rao, S.S.; Punnoose, D.; Tulasivarma, C.V.; Gopi, C.V.V.M.; Prabakar, K.; Kim, H.-J. Influence of solvents in the preparation of cobalt sulfide for supercapacitors. *R Soc. Open Sci.* **2017**, *4*, 170427. [CrossRef]
14. Lukatskaya, M.R.; Mashtalir, O.; Ren, C.E.; Agnese, Y.D.; Rozier, P.; Taberna, P.L.; Naguib, M.; Simon, P.; Barsoum, M.W.; Gogotsi, Y. Cation intercalation and high volumetric capacitance of two-dimensional titanium carbide. *Science* **2013**, *341*, 1502–1505. [CrossRef]
15. Guo, D.; Zhang, P.; Zhang, H.; Yu, X.; Zhu, J.; Li, Q.; Wang, T. NiMoO$_4$ nanowires supported on Ni foam as novel advanced electrodes for supercapacitors. *J. Mater. Chem. A* **2013**, *1*, 9024–9027. [CrossRef]

16. Kumar, Y.A.; Kim, H.-J. Effect of time on hierarchical corn skeleton-like composite of CoO@ZnO as capacitive electrode material for high specific performance supercapacitors. *Energies* **2018**, *11*, 3285. [CrossRef]

17. Yu, X.; Lu, B.; Xu, Z. Super long-life supercapacitors based on the construction of nanohoneycomb-like strongly coupled CoMoO$_4$-3D graphene hybrid electrodes. *Adv. Mater.* **2014**, *26*, 1044–1051. [CrossRef]

18. Wu, H.; Lou, Z.; Yang, H.; Shen, G. A flexible spiral-type supercapacitor based on ZnCo$_2$O$_4$ nanorod electrodes. *Nanoscale* **2015**, *7*, 1921–1926. [CrossRef] [PubMed]

19. Zhang, Q.; Deng, Y.; Hu, Z.; Liu, Y.; Yao, M.; Liu, P. Seaurchin-like hierarchical NiCo$_2$O$_4$@NiMoO$_4$ core-shell nanomaterials for high performance supercapacitors. *Phys. Chem. Chem. Phys.* **2014**, *16*, 23451–23460. [CrossRef] [PubMed]

20. Chen, H.; Jiang, J.; Zhang, L.; Qi, T.; Xia, D.; Wan, H. Facilely synthesized porous NiCo$_2$O$_4$ flowerlike nanostructure for high-rate supercapacitors. *J. Power Sources* **2014**, *248*, 28–36. [CrossRef]

21. Wei, T.Y.; Chen, C.H.; Chien, H.C.; Lu, S.Y.; Hu, C.C. A cost-effective supercapacitor material of ultrahigh specific capacitances: spinel nickel cobaltite aerogels from an epoxide-driven sol gel process. *Adv. Mater.* **2010**, *22*, 347–351. [CrossRef] [PubMed]

22. Cai, D.; Liu, B.; Wang, D.; Liu, Y.; Wang, L.; Li, H.; Wang, Y.; Wang, C.; Li, Q.; Wang, T. Enhanced performance of supercapacitors with ultrathin mesoporous NiMoO$_4$ nanosheets. *Electrochim. Acta* **2014**, *125*, 294–301. [CrossRef]

23. Jiang, Y.; Liu, B.; Zhai, Z.; Liu, X.; Yang, B.; Liu, L.; Jiang, X. A general strategy toward the rational synthesis of metal tungstate nanostructures using plasma electrolytic oxidation method. *Appl. Surf. Sci.* **2015**, *356*, 273–281. [CrossRef]

24. Zhang, G.; Lou, X.W. Controlled Growth of NiCo$_2$O$_4$ Nanorods and Ultrathin Nanosheets on Carbon Nanofibers for High-performance Supercapacitors. *Sci. Rep.* **2013**, *3*, 1470. [CrossRef] [PubMed]

25. Senthilkumar, B.; Sankar, K.V.; Selvan, R.K.; Danielle, M.; Manickam, M. Nano α-NiMoO$_4$ as a new electrode for electrochemical supercapacitors. *RSC Adv.* **2013**, *3*, 352. [CrossRef]

26. Shen, L.; Che, Q.; Li, H.; Zhang, X. Mesoporous NiCo2O4 nanowire arrays grown on carbon textiles as binder-free flexible electrodes for energy storage. *Adv. Funt. Mater.* **2014**, *24*, 2630–2637. [CrossRef]

27. Mai, L.-Q.; Yang, F.; Zhao, Y.-L.; Xu, X.; Xu, L.; Luo, Y.-Z. Hierarchical MnMoO$_4$/CoMoO$_4$ heterostructured nanowires with enhanced supercapacitor performance. *Nat. Commun.* **2011**, *2*, 381. [CrossRef] [PubMed]

28. Cui, C.; Xu, J.; Wang, L.; Guo, D.; Mao, M.; Ma, J.; Wang, T. Growth of NiCo2O4@MnMOO4 nanocolumn arrays with superior pseudocapacitor properties. *ACS Appl. Mater. Interfaces* **2016**, *8*, 8568–8575. [CrossRef]

29. Cheng, J.; Lu, Y.; Qiu, K.; Yan, H.; Xu, J.; Han, L.; Liu, X.; Luo, J.; Kim, J.-K.; Luo, Y. Hierarchical core/shell NiCo$_2$O$_4$@NiCo$_2$O$_4$ nanocactus arrays with dual functionalities for high performance supercapacitors and Li-ion batteries. *Sci. Rep.* **2015**, *5*, 12099. [CrossRef]

30. Xu, Z.W.; Li, Z.; Tan, X.H.; Holt, C.M.B.; Zhang, L.; Amirkhiz, B.S. Supercapacitive carbon nanotube – cobalt molybdate nanocomposites prepared via solvent free microwave synthesis. *RSC Adv.* **2012**, *2*, 2753–2755. [CrossRef]

31. Gobal, F.; Faraji, M. Preparation and electrochemical performances of nanoporous/cracked cobalt oxide layer for supercapacitors. *Appl. Phys. A* **2014**, *117*, 2087–20194. [CrossRef]

32. Meng, X.; Sun, H.; Zhu, J.; Bi, H.; Han, Q.; Liu, X.; Wang, X. Graphene-based cobalt sulfide composite hydrogel with enhanced electrochemical properties for supercapcitors. *New J. Chem.* **2016**, *40*, 2843–2849. [CrossRef]

33. Purushothaman, K.K.; Cuba, M.; Muralidharan, G. Supercapacitor behavior of α-MnMoO$_4$ nanorods on different electrolytes. *Mater. Res. Bull.* **2012**, *47*, 3348–3351. [CrossRef]

34. Park, K.S.; Seo, S.D.; Shim, H.W.; Kim, D.W. Electrochemical performance of Ni$_x$Co$_{1-x}$MoO$_4$ (0$\leqq$x$\leqq$1) nanocire anodes for lithium-ion batteries. *Nanoscale Res. Lett.* **2012**, *7*, 35–41. [CrossRef]

35. Xiao, W.; Chen, J.S.; Li, M.C.; Xu, R.; Lou, X.W. Synthesis, characterization, and lithium storage capability of AMoO$_4$ (A = Ni, Co) nanorods. *Chem. Mater.* **2010**, *22*, 746–754. [CrossRef]

36. Chavan, H.S.; Hou, B.; Ahmed, A.T.A.; Jo, Y.; Cho, S.; Kim, J.; Pawar, S.M.; Cha, S.N.; Inamdar, A.I.; Im, H.; Kim, H. Nanoflake NiMoO$_4$ based smart supercapacitor for intelligent power balance monitoring. *Sol. Energy Mater. Sol. Cells* **2018**, *185*, 166–173. [CrossRef]

37. Yedluri, A.K.; Kim, H.-J. Enhanced electrochemical performance of nanoplate nickel cobaltite (NiCo$_2$O$_4$) supercapacitor applications. *RSC Adv.* **2019**, *9*, 1115. [CrossRef]

38. Zhou, Y.; Song, J.; Li, H.; Feng, X.; Huang, Z.; Chen, S.; Ma, Y.; Wang, L.; Yan, X. The synthesis of shape-controlled α-MoO_3/graphene nanocomposites for high performance supercapacitors. *New J. Chem.* **2015**, *39*, 8780–8786. [CrossRef]

39. Lin, L.; Liu, T.; Liu, J.; Sun, R.; Hao, J.; Ji, K.; Wang, Z. Facile synthesis of groove-like $NiMoO_4$ hollow nanorods for high-performance supercapacitors. *Appl. Surface Sci.* **2016**, *360*, 234–239. [CrossRef]

40. Xia, X.F.; Lei, W.; Hao, Q.L. One-step synthesis of $CoMoO_4$/graphene composites with enhanced electrochemical properties for supercapacitors. *Electrochim. Acta* **2013**, *99*, 253–261. [CrossRef]

41. Yan, J.; Fan, Z.J.; Sun, W.; Ning, G.Q.; Wei, T.; Zhang, Q. Advanced asymmetric supercpacitors based on $Ni(OH)_2$/graphene and porous graphene electrodes with high energy density. *Adv. Funct. Mater.* **2012**, *22*, 2632. [CrossRef]

42. Roginskaya, Y.E.; Morozova, O.; Lubnin, E.; Ulitina, Y.E.; Lopukhova, G.; Trasatti, S. Characterization of bulk and surface composition of $Co_xNi_{1-x}O_y$ mixed oxides for electrocatalysis. *Langmuir* **1997**, *13*, 4621–4627. [CrossRef]

43. Marco, J.; Gancedo, J.; Gracia, M.; Gautier, J.; RÍos, E.; Berry, F.J. Characterization of the nickel cobaltite, prepared by several methods: an XRD, XANES, and XPS study. *Solid State Chem.* **2000**, *153*, 74–81. [CrossRef]

44. Yao, M.; Hu, Z.; Liu, Y.; Liu, P. Design and synthesis of hierarchical $NiCo_2S_4$@$NiMoO_4$ core/shell nanospheres for high-performance supercapacitors. *New J. Chem.* **2015**, *39*, 8430. [CrossRef]

45. Hong, W.; Wang, J.Q.; Gong, P.W.; Sun, J.F.; Niu, L.Y.; Yang, Z.G.; Wang, Z.F.; Yang, S.R. Rational construction of three dimensional hybrid Co_3O_4@$NiMoO_4$ nanosheets array for energy storage application. *J. Power Sources* **2014**, *270*, 516. [CrossRef]

46. Liu, M.C.; Kong, L.B.; Lu, C.; Ma, X.J.; Li, X.M.; Luo, Y.C.; Kang, L. Design and synthesis of $CoMoO_4$-$NiMoO_4$.xH_2O bundles with improved electrochemical properties for supercapacitors. *J. Mater. Chem. A* **2013**, *1*, 1380. [CrossRef]

47. Cai, D.P.; Wang, D.D.; Liu, B.; Wang, L.L.; Liu, Y.; Li, H.; Wang, Y.R.; Li, Q.H.; Wang, T.H. Three dimensional Co_3O_4@$NiMoO_4$ core/shell nanowire arrays on Ni foam for electrochemical energy storage. *ACS Appl. Mater. Interfaces* **2014**, *6*, 5050–5055. [CrossRef] [PubMed]

48. Zhang, Z.; Liu, Y.; Huang, Z.; Ren, L.; Qi, X.; Wei, X.; Zhong, J. Facile hydrothermal synthesis of $NiMoO_4$@$CoMoO_4$ hierarchical nanospheres for supercapacitor applications. *J. Phys. Chem. Phys.* **2015**, *17*, 20795–20804. [CrossRef] [PubMed]

49. Huang, L.; Zhang, W.; Xiang, J.; Xu, H.; Li, G.; Huang, Y. Hierarchical core-shell $NiCo_2O_4$@$NiMoO_4$ nanowires grown on carbon cloth as integrated electrode for high-performance supercapacitors. *Sci. Rep.* **2016**, *6*, 31465. [CrossRef]

Article

Facilely Synthesized NiCo$_2$O$_4$/NiCo$_2$O$_4$ Nanofile Arrays Supported on Nickel Foam by a Hydrothermal Method and Their Excellent Performance for High-Rate Supercapacitance

Anil Kumar Yedluri [iD], **Eswar Reddy Araveeti and Hee-Je Kim** *

School of Electrical Engineering, Pusan National University, Busandaehak-ro 63beon-gil, Geumjeong-gu, Busan 46241, Korea; yedluri.anil@gmail.com (A.K.Y.); easwarreddy.a@gmail.com (E.R.A.)
* Correspondence: heeje@pusan.ac.kr; Tel.: +82-10-2295-0613; Fax: +82-51-513-0212

Received: 2 February 2019; Accepted: 2 April 2019; Published: 5 April 2019

Abstract: NiCo$_2$O$_4$ nanoleaf arrays (NCO NLAs) and NiCo$_2$O$_4$/NiCO$_2$O$_4$ nanofile arrays (NCO/NCO NFAs) material was fabricated on flexible nickel foam (NF) using a facile hydrothermal approach. The electrochemical performance, including the specific capacitance, charge/discharge cycles, and lifecycle of the material after the hydrothermal treatment, was assessed. The morphological and structural behaviors of the NF@NCO NLAs and NF@NCO/NCO NFAs electrodes were analyzed using a range of analysis techniques. The as-obtained nanocomposite of the NF@NCO/NCO NFAs material delivered outstanding electrochemical performance, including an ultrahigh specific capacitance (Cs) of 2312 F g^{-1} at a current density of 2 mA cm^{-2}, along with excellent cycling stability (98.7% capacitance retention after 5000 cycles at 5 mA cm^{-2}). These values were higher than those of NF@NCO NLAs (Cs of 1950 F g^{-1} and 96.3% retention). The enhanced specific capacitance was attributed to the large electrochemical surface area, which allows for higher electrical conductivity and rapid transport between the electrons and ions as well as a much lower charge-transfer resistance and superior rate capability. These results clearly show that a combination of two types of binary metal oxides could be favorable for improving electrochemical performance and is expected to play a major role in the future development of nanofile-like composites (NF@NCO/NCO NFAs) for supercapacitor applications.

Keywords: nickel foam (NF); NiCo$_2$O$_4$ nanoleaf; NiCo$_2$O$_4$/NiCo$_2$O$_4$ nanofile; Superior electrochemical performance; hydrothermal approach; supercapacitor

---

1. Introduction

One of the greatest scientific and technological societal challenges of the present time is achieving secure, sustainable, and highly effective energy conversion and storage equipment system [1–3]. Over the last few decades, a range of energy storage devices, such as solar cells, batteries, lithium-ion batteries, electrochemical capacitors, fuel cells, and supercapacitors (SCs), have been reported [4]. Among the above energy storage devices, SCs have considerable potential and are used widely in storage devices owing to their rapid charge-discharge cycle performance, outstanding cyclability, eco-friendly nature, long cycling life span, and high rate abilities [5–7]. Various materials for faradic redox reactions (pseudocapacitors (PCs), such as Co$_3$O$_4$, NiO, MnO$_2$, and RuO$_2$, have been used as electrodes to achieve high energy capability for SCs [8,9]. Nevertheless, the lower electrical conductivity of traditional PCs limits their utility because they cannot provide high performance at high current densities. Recently, new types of bimetallic oxides (BMOs), such as CuCo$_2$O$_4$, CoMoO$_4$, NiMoO$_4$, MnMoO$_4$, and NiCo$_2$O$_4$, have attracted considerable attention in the field of SCs because of their high energy storage performance, including high electrical conductivity and superior energy density compared to common PC oxides [10–12]. Therefore, it is important to develop various bimetallic oxide materials for higher performance SCs.

Among the BMOs available, $NiCo_2O_4$ has attracted attention for use in high performance electrochemical SCs applications owing to its outstanding electrochemical conductivity, high theoretical capacitance, and rapid redox activities [13–15]. Different types of nanostructures ($NiCo_2O_4$), including nanoplates/sheets, nanoparticles/wires and needles, nanospheres/tubes, and nanorods, have been fabricated for SC applications [16–18]. For example, Zou et al. fabricated chain-like $NiCo_2O_4$ nanowires using a facile hydrothermal and thermal decomposition approach and reported a specific capacitance (Cs) of 1284 F g^{-1} at 2 A g^{-1} [19]. Zhang et al. synthesized hierarchical mesoporous spinal $NiCo_2O_4$ via a simple hydrothermal and post annealing process and reported a Cs of 1619.1 F g^{-1} at a 2.0 A g^{-1} [20]. Lei et al. synthesized three-dimensional (3D) hierarchical flower-shaped $NiCo_2O_4$ microspheres through a microwave-assisted heating reflux method and a delivered a Cs of 1006 F g^{-1} at 1 A g^{-1} [21]. Moreover, Yedluri et al. fabricated $NiCo_2O_4$ nanoplates by a simple chemical bath deposition approach and reported a Cs of 1018 F g^{-1} at 20 mA g^{-2} [22]. Recently, Yuan et al. synthesized ultrathin mesoporous $NiCo_2O_4$ nanosheets through a co-electrodeposition approach and reported a Cs of 1450 F g^{-1} at 20 A g^{-1} [23]. Zhu et al. fabricated a spinal $NiCo_2O_4$ electrode by facile hydrothermal synthesis and achieved a high Cs of 1254 F g^{-1} at 2 A g^{-1} [24]. Zou et al. synthesized $NiCo_2O_4$ nanowires through a hydrothermal approach and reported a high Cs of 1284 F g^{-1} at 2 A g^{-1} [25]. Despite these studies highlighting the advantages of $NiCo_2O_4$ electrodes, their poor specific capacitance still limits their wider commercialization; the very low energy and power densities are challenges that remain to be solved.

The use of active materials in SCs should be improved to increase the specific capacitance, cycling stability, and energy storage. Combining two types of BMOs to form unique hierarchical nanostructures is an efficient way to improve the electrochemical (EC) performance. For example, Cui et al. prepared $NiCo_2O_4@MnMoO_4$ nanocolumn array structures via a two-step hydrothermal route, which delivered a Cs of 1705.3 F g^{-1} at 5 mA cm^{-2} [26]. Mai et al. synthesized $MnMoO_4@CoMoO_4$ nanowires via a facile hydrothermal method and achieved a high Cs of 187.1 F g^{-1} at 1 A g^{-1} [27]. In addition, Yuan et al. used a facile two-step hydrothermal route to develop a $NiCo_2O_4@MnMoO_4$ core-shell nanoarray skeleton, exhibiting a high Cs of 1169 F g^{-1} at 2.5 mA cm^{-2} [28]. Furthermore, Cheng et al. reported a core/shell $NiCo_2O_4@NiCo_2O_4$ nanocactus for SC applications, showing a super high Cs of 1264 F g^{-1} at 2 A g^{-1} [29]. Recently, Gu et al. used a facile hydrothermal method to synthesize core-shell $NiCo_2O_4@MnMoO_4$ nanomaterials over a nickel foam substrate and reported a high Cs of 1118 F g^{-1} at 1 A g^{-1} and high cycling stability [30]. Zhang et al. presented hierarchical nanospheres of a $NiMoO_4@CoMoO_4$ electrode material via a hydrothermal method for SCs applications, showing an excellent Cs of 1601.6 F g^{-1} at 2 A g^{-1} with superior cycling stability [31]. Overall, these studies suggested that a combination of two types of BMOs material is crucial for improving the high EC performance. To the best of our knowledge, the performance of an NF@NCO nanoleaf arrays (NLAs) and NF@NCO/NCO nanofile arrays (NFAs) composite is not satisfactory, and it is a big challenge to further improve the energy storage performance of SCs using NF@NCO NLAs and NF@NCO/NCO NFAs composite materials. Herein, we presented the synthesis of NF@NCO NLAs and NF@NCO/NCO NFAs composite nanostructures using a facile hydrothermal approach. Owing to the high surface area and structural features, the NF@NCO NLAs and NF@NCO/NCO NFAs composite electrodes delivered high specific capacitances with good cycling stabilities when employed as an electrode material for SCs. Compared to the previously reported composite metal oxides, the excellent electrochemical results suggest that the NF@NCO/NCO NFAs composite could be a promising candidate for future research work on SCs applications.

Inspired by the above ideas, this study fabricated a hierarchical NCO/NCO NFAs material on a nickel foam (NF) substrate via a facile hydrothermal technique for high-performance SC applications. Compared to the above electrochemical (EC) performances, the hierarchical NF@NCO/NCO NFAs electrode material provided a facile synthetic process to simultaneously enhance the specific capacitance (Cs), cycling stability, and electrical conductivity toward flexible high-performance EC applications. The NF@NCO/NCO NFAs scaffold architecture, which grows on another network SC of an NF@NCO NLAs material, provides a large surface area with extensive electrochemical active sites. The NF@NCO/NCO NFAs electrode delivered a high Cs of 2312 F g^{-1} at 2 mA cm^{-2} and excellent cycling stability (98.7% capacitance retention after 5000 cycles) compared to SCs of NF@NCO NLAs (Cs of 1950 F g^{-1} at 2 mA cm^{-2} and 96.3% capacitance retention after 5000 cycles). This type of combination can be expanded to a range of BMOs for various integrated devices to enhance the Cs, cycling life span, and energy storage performance for improved applications.

2. Experimental Methods

2.1. Chemicals and Materials

Analytical grade nickel nitrate hexahydrate (Ni(NO$_3$)$_2$·6H$_2$O), cobalt chloride hexahydrate (CoCl$_2$·6H$_2$O), thiourea (CH$_4$N$_2$S), hexamethylenetetramine (C$_6$H$_{12}$N$_4$), and potassium hydroxide (KOH) were obtained from Sigma-Aldrich Co. manufacturer, Seoul city, South Korea and used directly in the fabrication of NF@NCO/NCO NFAs.

2.2. Fabrication of the NF@NCO NLAs and NF@NCO/NCO NFAs Materials

The NF@NCO/NCO NFAs and NF@NCO NLAs electrode materials used a simple hydrothermal technique. Before the growth process, the NF (approximately 1.5 × 2 cm^2) was cleaned sequentially with acetone, 2 M HCl solution, ethanol, and deionized water (DI) with ultrasonication for 20 min each and then dried at 60 °C for 8 h. In a typical hydrothermal procedure, the NF@NiCo$_2$O$_4$ NLAs were prepared in a round bottom flask containing a 100 mL of solution of 0.1 M of Ni(NO$_3$)$_2$·6H$_2$O and 0.1 M of Co(NO$_3$)$_2$·6H$_2$O in DI water with vigorous stirring for 1 h using a magnetic stirrer. Subsequently, 0.8 M of CH$_4$N$_2$S and 0.08 M of C$_6$H$_{12}$N$_4$ were added slowly to the previous solution with stirring for 2 h. Finally, a clear pink solution was formed under vigorous stirring. The prepared recipe was then transferred to a 100 mL Teflon-lined stainless-steel (TLSS) autoclave with the prepared NF as growth solution and heated at 160 °C for 18 h. After the reaction, the NF@NCO NLAs product electrodes were removed from the TLSS autoclave and cooled naturally to room temperature. The resulting product was then washed with DI water and ethanol several times. Finally, the obtained product was dried at 100 °C for 10 h to obtain the NF@NCO NLAs.

Finally, the NF@NCO/NCO NFAs composite was fabricated using the following process. A NF@NCO/NCO NFAs electrode was also prepared using the above relevant chemicals for the NF@NCO NLAs, including 0.1 M of Ni(NO$_3$)$_2$·6H$_2$O, 0.1 M of Co(NO$_3$)$_2$·6H$_2$O, 0.8 M of CH$_4$N$_2$S, and 0.08 M of C$_6$H$_{12}$N$_4$. The as-prepared solution was vigorously stirred, and then NF@NCO NFAs foams were placed vertically in the prepared solution, kept at 150 °C for 20 h in an electric oven, then cooled naturally to room temperature. After the reaction, the obtained Ni foams were removed from the oven and washed sequentially with DI water and ethanol and then dried at overnight at 90 °C in order to obtain the final composite (NF@NCO/NCO NFAs) material. The mass-loading of the active material in NF@NCO/NCO NFAs and NF@NCO NLAs were calculated to be 6 mg cm^{-2} and 5 mg cm^{-2}, respectively.

2.3. Material Characterizations

The fabricated NF@NCO NLAs and NF@NCO/NCO NFAs samples were analyzed by field emission scanning electron microscopy (HR-SEM, Hitachi S4800) to study their microstructure and surface morphology. High-resolution transmission electron microscopy (HR-TEM, CJ111) was performed to investigate the crystalline structure and morphology. X-ray photoelectron spectroscopy (XPS, ESCALAB-250Xi) was conducted to examine the chemical composition and surface states.

X-ray diffraction (XRD, D/max-2400, Rigaku at 30 kV and 30 mA) was performed to study the crystal structure.

2.4. Electrochemical Measurements

Cyclic voltammetry (CV), galvanostatic charge-discharge (GCD) analysis, and electrochemical impedance spectroscopy (EIS) of the as-developed electrodes (NF@NCO NLAs and NF@NCO/NCO NFAs) were performed using a Biologic SP-150 electrochemical workstation in a three-electrode setup with a 2.0 M KOH aqueous electrolyte. In the three-electrode setup, the as-developed samples, platinum wire, and Ag/AgCl were used as the working, counter, and reference electrodes, respectively. All electrodes were tested in a 2.0 M KOH electrolyte with CV recorded at the potential window of -0.2 to 0.5 V (V versus Ag/AgCl). GCD was recorded at 0 to 0.4 V versus Ag/AgCl. For the three-electrode setup, the specific capacitance (C_s), energy density (E, W h kg^{-1}), and power density (P, W kg^{-1}) were calculated from the GCD curves using following equations:

$$C_s = \frac{I \times t}{m \times V} \tag{1}$$

$$E = \frac{C_s \times (\Delta V^2)}{2} \tag{2}$$

$$P = \frac{E}{t} \tag{3}$$

where I (A) represents current density; m (g) is the mass of the electrode; ΔV (V) is the voltage window; and Δt (s) is the discharge time difference [32].

3. Results and Discussion

Figure S1 in the Supplementary material presents schematic illustration of NF@NCO/NCO NFAs. Figure 1a–f presents FESEM images of the NF@NCO/NCO NFAs and NF@NCO NLAs samples. Figure 1a–c depicts the SEM images of the NF@NCO/NCO NFAs on NF substrate at various magnifications. Leaf-like NF@NCO NLAs were grown compactly over the surface of NF@NCO/NCO NFAs. The low-magnification SEM images of NF@NCO/NCO NFAs (Figure 1a–c) showed that the NF was covered uniformly and densely with the as-deposited material. As shown in Figure 1b, the nanofiles were interconnected with each other to form a stable and uniform structure with a network structure and no voids on the NF surface area. The high-magnification image in Figure 1c revealed highly ordered nanofiles grown homogeneously over the surface to form appropriate diffusion channels for electrolyte penetration, which provides a large specific surface area that boosts the intercalation/deintercalation rate, undoubtedly enhances the electrochemical performance, and enables a high transportation efficiency of ions and electrons. FESEM of the as-fabricated leaf-like structure of NF@NCO NLAs at various magnifications revealed a different structure (Figure 1d–f). NF@NCO possesses a type of wire/leaf-like nanostructure morphology with various leaf sizes, a large number of voids, and loosely connected networks surrounded by the surface of the nanoleaf (Figure 1d,e). This type of network delivered weak intercalation between the NCO and NF surface area and resulted in poorer electrochemical performance with lower electrical conductivity, as shown in Figure 1f.

Figure 2 presents HR-TEM, TEM, and HAADF-STEM with EDX images of NF@NCO/NCO NFAs and NF@NCO NLAs. The structure and morphology of the active NF@NCO/NCO NFAs electrodes were analyzed, and are shown in Figure 2a,b. The nanofile morphology was observed over the entire surface of NF@NCO NLAs and the results concur with the FE-SEM images (Figure 1a–c). The HR-TEM image (Figure 2b) showed that the lattice fringe spacing of the files were 0.247 and 0.205 nm, which were assigned to the (311) and (400) planes of the file-like nanostructured NiCO$_2$O$_4$, respectively. As shown in Figure 2c,d, the HR-TEM and TEM of the NF@NCO NLAs sample taken from the NF substrate clearly showed a flake-like nanostructure similar to FESEM (Figure 1d–f). In addition, the HAADF-STEM with EDS mapping analysis revealed a homogeneous distribution and the coexistence of Co, Ni, and O in the NF@NCO/NCO NFAs electrode (Figure 2e,f).

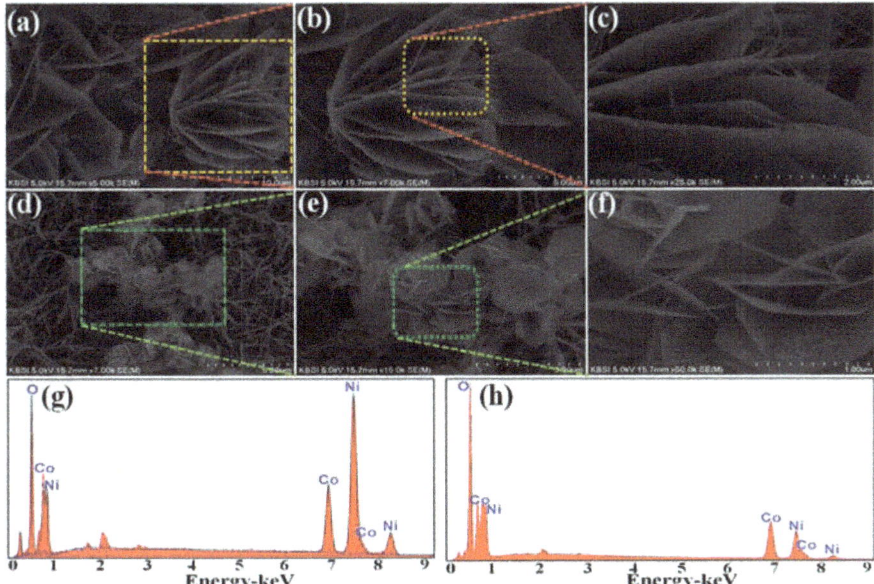

Figure 1. Morphological characterization of NF@NCO/NCO nanofile arrays (NFAs) and NF@NCO nanoleaf arrays (NLAs). FESEM images of the prepared NF@NCO/NCO NFAs (**a–c**) and NF@NCO NLAs (**d–f**) and their corresponding high magnification images and EDX data (**g**) and (**h**).

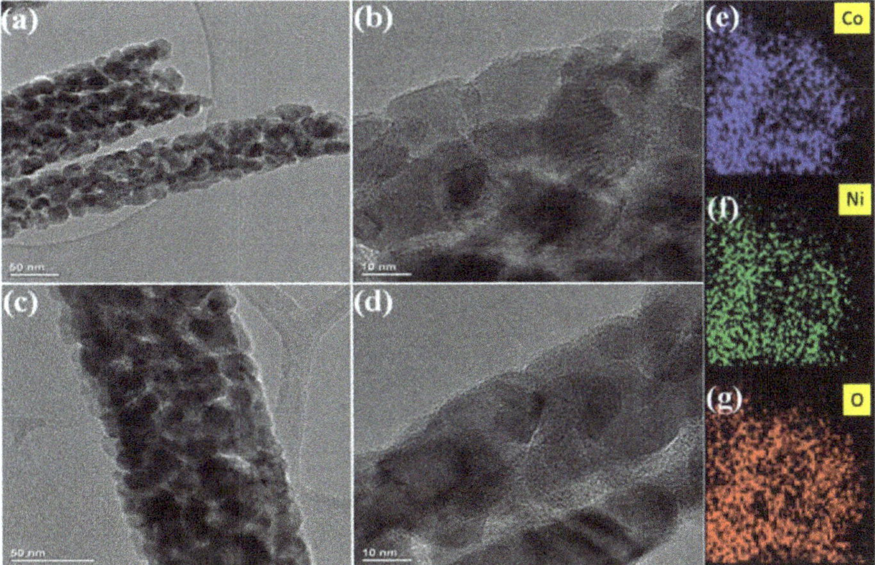

Figure 2. TEM and HR-TEM images of the NF@NCO/NCO NFAs (**a**) and (**b**) and NF@NCO NLAs (**c**) and (**d**) grown on nickel foam at different magnifications and STEM-EDS color elemental mapping images of the NF@NCO/NCO NFAs electrode (**e–g**).

The fabricated NF@NCO/NCO NFAs and NF@NCO NLAs samples were examined further by powder XRD and XPS to determine the crystal nature and phase purity. Figure 3a shows the XRD patterns of the NF@NCO/NCO NFAs, NF@NCO NLAs, and NF substrate samples, respectively. As shown in Figure 3a, XRD peaks were observed at (111), (220), (222), (422), and (511), and were attributed to the spinal cubic structure of $NiCo_2O_4$ and well-matched with the JCPDS card number (JCPDS card No. 73-1702) [33]. Other XRD peaks at (311), (400), and (440) were also observed due to the NF substrate and were well-matched with the JCPDS card number (JCPDS card No-04-0850). No other impurity or other peaks were observed, which confirmed the purity of the as-fabricated material.

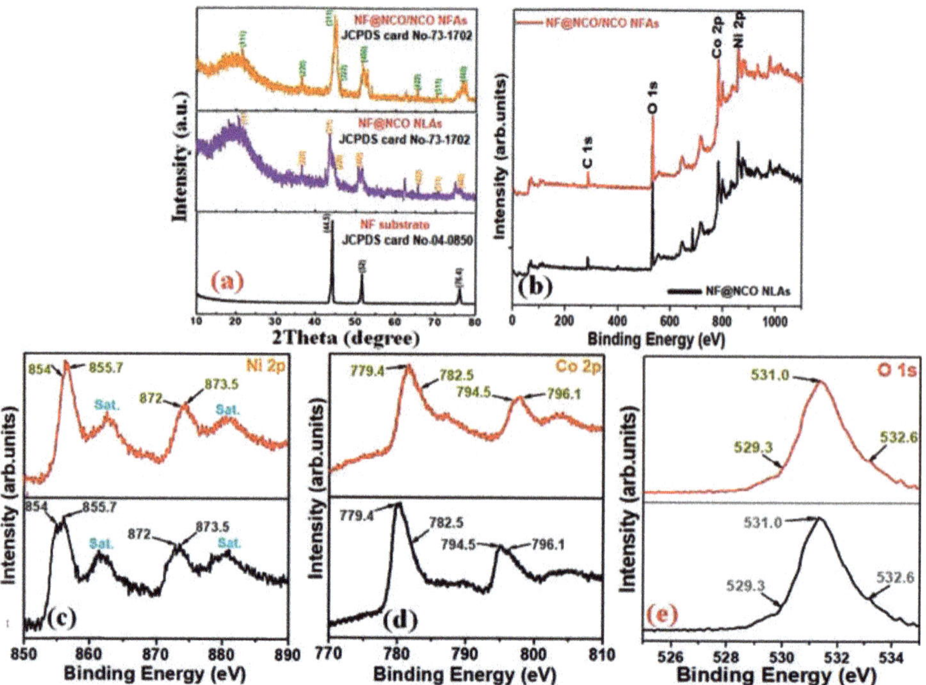

Figure 3. (a) XRD pattern of the as-synthesized NF@NCO/NCO NFAs, NF@NCO NLAs, and Nickel foam material, (b) XPS survey spectrum of NF@NCO NLAs and NF@NCO/NCO NFAs, and (c–e) high-resolution XPS spectrum of Ni 2p (c), Co 2p, and O1s (e) for NF@NCO/NCO NFAs and NF@NCO NLAs.

The oxidation states and elemental identification of NF@NCO/NCO NFAs and NF@NCO NLAs electrodes were examined by XPS, as shown in Figure 3b. Figure 3b shows the survey spectrum of both samples. Ni 2p, Co 2p, and O1s signals with no other peaks were detected, which confirmed that the samples were pure. Figure 3c shows the convoluted Ni 2p spectrum, which exhibits two major peaks and two broad satellites (Sat.), such as Ni $2p_{3/2}$ peaks at approximately 854.0 and 872.0 eV indicating Ni^{2+} and Ni^{3+}, respectively, and Ni $2p_{1/2}$ peaks at 855.7 and 873.5 eV for Ni^{2+} and Ni^{3+}, respectively [34]. The high-resolution Co 2p spectrum (Figure 3d) revealed peaks at 779.4 and 794.5 eV, which were assigned to Co^{2+}, and the Co 2p peaks at 782.5 and 796.1 eV were attributed to Co^{3+}, which are in agreement with previous studies. Figure 3e [35] presents the O 1s spectrum, in which there are three main peaks that appear at 529.3, 531.0, and 532.6 eV, corresponding to metal-oxygen bonds and the multiplicity of chemisorbed oxygen-metal bonds within the surface of the samples (NF@NCO/NCO NFAs and NF@NCO NLAs) [36–38].

The electrochemical properties of the NF@NCO NLAs and NF@NCO/NCO NFAs samples were investigated in a three-electrode system with an aqueous 2.0 M KOH electrolyte using CV and GCD techniques. Figure 4a shows the CV curves of the as-prepared electrodes at a scan rate of 50 mV s^{-1} over the voltage window from -0.2 to 0.5 V, respectively. The CV integrated area and the oxidation and reduction currents of the NF@NCO/NCO NFAs electrode were much larger than that of the NF@NCO NLAs electrode. In addition, the electrocapacitance of both as-prepared electrodes was examined by GCD measurements at 2 mA cm^{-2} in a voltage window of 0 to 0.4 V, as shown in Figure 4b. The GCD plots revealed significant differences between the two as-prepared electrodes, which showed a larger enclosed area for the NF@NCO/NCO NFAs than that of the NF@NCO NLAs electrode.

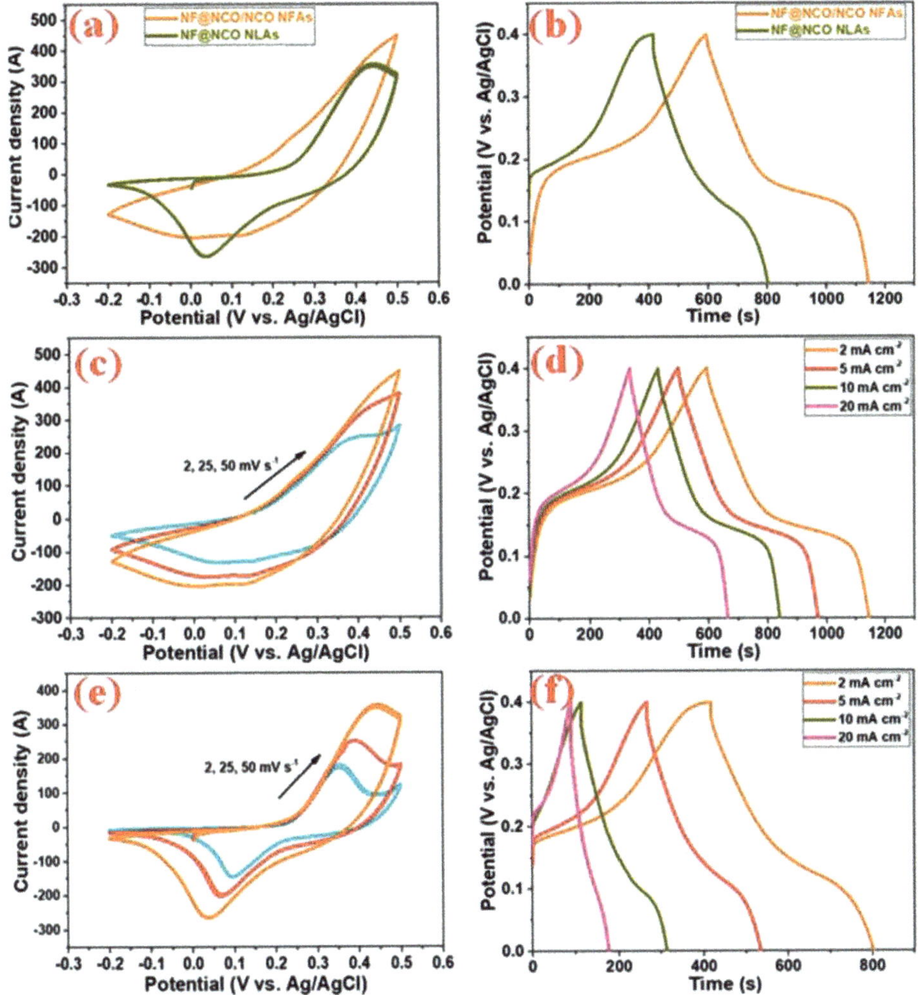

Figure 4. Comparison of (**a**) cyclic voltammetry (CV) and (**b**) galvanostatic charge-discharge (GCD) plots of the NF@NCO NLAs and NF@NCO/NCO NFAs electrodes measured at a constant scan rate and current density. (**c**) CV and (**d**) GCD behaviors of the NF@NCO/NCO NFAs electrodes at various scan rates (2, 25, and 50 mV s^{-1}) and current densities (2–20 mA cm^{-2}). (**e**) CV and (**f**) GCD behaviors of the NF@NCO NLAs electrodes at various scan rates (2, 25, and 50 mV s^{-1}), and current densities (2–20 mA cm^{-2}).

Figure 4c,e present the CV curves of the as-prepared electrodes (NF@NCO/NCO NFAs and NF@NCO NLAs) analyzed over the voltage window, -0.2 to 0.5 V, at scan rates of 2, 25, and 50 mV s^{-1}. Typical CV curves (Figure 4c,e) of the NF@NCO/NCO NFAs composite electrode material showed a much larger integrated area than that of the NF@NCO NLAs electrode, indicating good ion diffusion, good reversibility of active materials, and high specific capacitance. In addition, the anodic and cathodic peak position shifted to lower and higher potentials, respectively, and exhibited a similar structure, which is an indication of the kinetic reversibility and pseudocapacitance behavior of the as-prepared electrodes. The GCD curves of the NF@NCO/NCO NFAs and NF@NCO NLAs electrodes were analyzed at current densities ranging from 2 to 20 mA cm^{-2}, as shown in Figure 4d,f. The GCD curves of the two as-obtained electrodes were similar, and a pair of charge-discharge plots that were similar to the reflection process of oxidation and reduction were obtained. Moreover, the flat voltage plateaus in the GCD results indicate typical pseudocapacitive behavior, which is in agreement with the electrochemical characteristics of the CV curves.

The specific capacitance of the NF@NCO NLAs and NF@NCO/NCO NFAs samples at current densities of 2, 5, 10, and 20 mA cm^{-2} were 2312, 2172, 2030, and 1782 F g^{-1}, respectively. In contrast, the corresponding specific capacitance of the NF@NCO NLAs electrodes were 1950, 1595, 1328, and 715 F g^{-1}, respectively. These results confirm that the NF@NCO/NCO NFAs electrode had a much higher specific capacitance than the NF@NCO NLAs electrode. The remarkable electrochemical performance of the NF@NCO/NCO NFAs electrode could be attributed to the excellent adhesion to the NF substrate with a large surface area and electrical connection of the active material to the current collector to ensure effective accessibility of the electrolyte ions and electrons. Moreover, EIS was conducted to examine the ion diffusion and electrical conductivity of the as-obtained samples, as depicted in Figure 5b. The measurements were performed over the frequency range 0.01 Hz to 100 kHz using a three-electrode system. The NF@NCO/NCO NFAs composite electrode exhibited a semicircle part in the high frequency region and a vertical line in the low frequency region, which represents the charge transfer resistance (R_{ct}) and equivalent series resistance (R_s), respectively. The nanofile-like NF@NCO/NCO NFAs electrode exhibited a lower R_{ct} (0.25 Ω) than the NF@NCO NLAs (R_{ct} = 0.5 Ω) electrode. Moreover, the NF@NCO/NCO NFAs electrode delivered a more vertical line than that of the NF@NCO NLAs electrode, indicating a higher efficient electro-active surface area that results in faster electrolyte ion diffusion during the redox reaction kinetics with excellent conductivity. In addition, the NF@NCO NLAs and NF@NCO/NCO NFAs supercapacitor devices exhibited an excellent stable high capacitance retention of 98.7% and 96.3%, respectively, after 5000 cycles at a current density of 5 mA cm^{-2} in a 2.0 M KOH aqueous electrolyte solution. Nevertheless, the NF@NCO/NCO NFAs supercapacitor device has more active sites and numerous pathways for electrolyte ion penetration, which promotes a rapid electrochemical reaction and improves the energy storage performance.

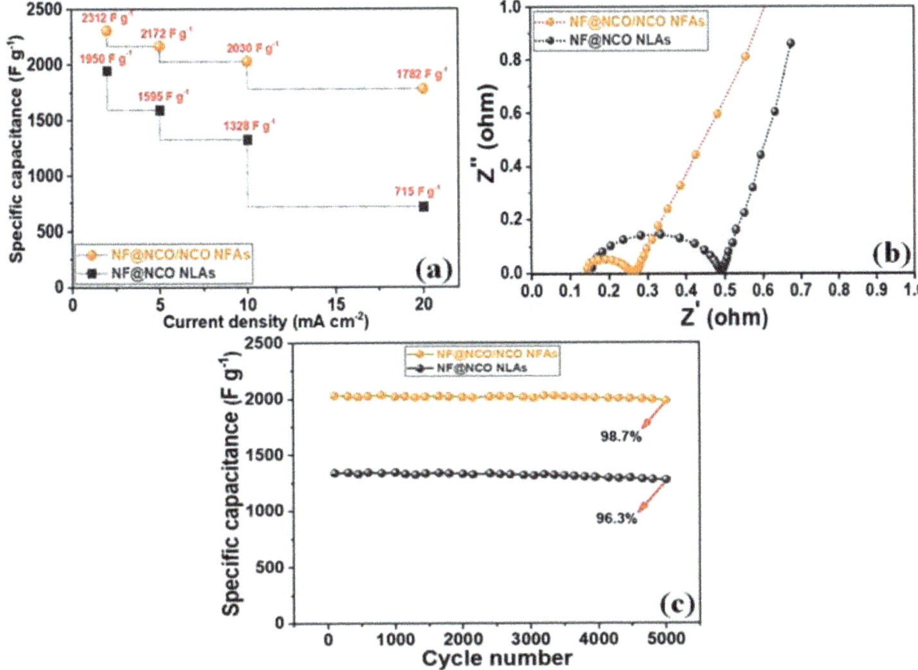

Figure 5. (**a**) Specific capacitance at various current densities, (**b**) Nyquist plots, and (**c**) Cycling stability of the as-prepared NF@NCO/NCO NFAs and NF@NCO NLAs samples.

4. Conclusions

A novel composite material with a NF@NCO/NCO NFAs-like structure was fabricated on an NF substrate using a facile hydrothermal process. The excellent supercapacitance performance of the NF@NCO/NCO NFAs was found to depend mainly on easy pathways between the transportation of electrons and ions, a large electrochemical surface area, more accessible electroactive sites, and great adhesion to the NF substrate to enhance energy storage and exhibit higher electrochemical performance than the NF@NCO NLAs electrode. The nanofile-like NF@NCO/NCO NFAs covered the NF@NCO NLAs surface area fully. As a result, the synthesized hierarchical nanofile-like NF@NCO/NCO NFAs composite electrode exhibited superior energy storage performance, including a high specific capacitance of 2312 F g^{-1} at a current density of 2 mA cm^{-2}, as well as a higher cycling stability at a current density of 5 mA cm^{-2} (98.7% retention after 5000 cycles) than the NF@NCO NLAs (96.3% retention after 5000 cycles). These results demonstrate the outstanding cycling stability, excellent electrochemical performance, and higher electrical conductivity of the hierarchical nanofile-like NF@NCO/NCO NFAs composite compared to the NF@NCO NLAs composite, highlighting its potential for flexible energy storage systems.

Supplementary Materials: The following are available online at http://www.mdpi.com/1996-1073/12/7/1308/s1, Figure S1: Schematic illustration of NF@NCO/NCO NFAs.

Author Contributions: Conceptualization, A.K.Y.; Formal analysis, H.-J.K. and E.R.A.; Investigation, A.K.Y. and H.-J.K.; Methodology, A.K.Y.; Supervision, A.K.Y.; Writing—original draft, A.K.Y.; Writing—review & editing, A.K.Y.

Funding: This research received no external funding.

Acknowledgments: This work was financially supported by BK 21 PLUS, the Creative Human Resource Development Program for IT Convergence (NRF-2015R1A4A1041584), Pusan National University, Busan, South Korea. We would like to thank KBSI, Busan for the SEM, TEM, XRD, XPS, and EDX analyses.

Conflicts of Interest: The authors declare no conflict of interest.

References

1. Gobal, F.; Faraji, M. Preparation and electrochemical performances of nanoporous/cracked cobalt oxide layer for supercapacitors. *Appl. Phys. A* **2014**, *117*, 2087–2094. [CrossRef]
2. Yang, M.; Choi, B.G. Preparation of Three-Dimensional Co_3O_4/Graphene Composite for High-Performance Supercapacitors. *Chem. Eng. Commun.* **2017**, *204*, 723–728. [CrossRef]
3. Le, Z.; Liu, F.; Nie, P.; Li, X.; Liu, X.; Bian, Z.; Chen, G.; Wu, H.B.; Lu, Y. Pseudocapacitive sodium storage in mesoporous single-crystalline-like TiO_2-Graphen nanocomposite enables high-performance sodium-ion capacitors. *ACS Nano* **2017**, *11*, 2952–2960. [CrossRef]
4. Augustyn, V.; Simon, P.; Dunn, B. Pseudocapacitive oxide materials for high-rate electrochemical energy storage. *Energy Environ. Sci.* **2014**, *7*, 1597–1614. [CrossRef]
5. Kumar, Y.A.; Rao, S.S.; Punnoose, D.; Tulasivarma, C.V.; Gopi, C.V.V.M.; Prabakar, K.; Kim, H.-J. Influence of solvents in the preparation of cobalt sulfide for supercapacitors. *R. Soc. Open Sci.* **2017**, *4*, 170427. [CrossRef]
6. Kumar, Y.A.; Kim, H.-J. Effect of Times on a Hierarchical Corn Skeleton-Like Composite of CoO@ZnO as Capacitive Electrode Material for High Specific Performance Supercapacitors. *Energies* **2018**, *11*, 3285. [CrossRef]
7. Singh, A.K.; Sarkar, D.; Karmakar, K.; Mandal, K.; Khan, G.G. High-Performance Supercapacitor Electrode based on Cobalt Oxide—Manganese Dioxide-Nickel Oxide Ternary 1D Hybrid Nanotubes. *ACS Appl. Mater. Interfaces* **2016**, *8*, 20786–20792. [CrossRef] [PubMed]
8. Younis, A.; Chu, D.; Li, S. Ethanol directed morphological evolution of hierarchical CeOx architectures as advance electrochemical capacitors. *J. Mater. Chem. A* **2015**, *3*, 13970–13977. [CrossRef]
9. Kumar, Y.A.; Kim, H.-J. Preparation and electrochemical performance of $NiCo_2O_4$@$NiCo_2O_4$ composite nanoplates for high performance supercapacitor applications. *New J. Chem.* **2018**, *42*, 19971–19978. [CrossRef]
10. Liu, S.; Mao, C.; Niu, Y.; Yi, F.; Hou, J.; Lu, S.; Jiang, J.; Xu, M.; Li, C.M. Facile Synthesis of Novel Networked Ultralong Cobalt Sulfide Nanotubes and Its Application in Supercapacitors. *ACS Appl. Mater. Interfaces* **2015**, *7*, 25568–25573. [CrossRef]
11. Xu, J.; Liu, S. Co_3O_4/ZnO nanoheterostructure derived from core–shell ZIF-8@ZIF-67 for supercapacitors. *RSC Adv.* **2016**, *6*, 52137–52142. [CrossRef]
12. Yedluri, A.K.; Kim, H.-J. Wearable super-high specific performance supercapacitors using a honeycomb with folded silk-like composite of $NiCo_2O_4$ nanoplates decorated with $NiMoO_4$ honeycombs on nickel foam. *Dalton Trans.* **2018**, *47*, 15545–15554. [CrossRef]
13. Bai, J.; Wang, K.; Feng, J.; Xiong, S. ZnO/CoO and $ZnCo_2O_4$ Hierarchical Bipyramid Nanoframes: Morphology Control, Formation Mechanism, and Their Lithium Storage Properties. *ACS Appl. Mater. Interfaces* **2015**, *7*, 22848–22857. [CrossRef] [PubMed]
14. Huang, K.-J.; Zhang, J.Z.; Shi, G.W.; Liu, Y.-M. One step hydrothermal synthesis of two-dimentional cobalt sulfide for high performance supercapacitors. *Mater. Lett.* **2014**, *131*, 45–48. [CrossRef]
15. Jiang, J.; Shi, W.; Song, S.; Hao, Q.; Fan, W.; Xia, X.; Zhang, X.; Wang, Q.; Liu, C.; Yan, D. Solvothermal synthesis and electrochemical performance in super-capacitors of Co_3O_4/C flower-like nanostructures. *J. Power Sources* **2014**, *248*, 1281–1289. [CrossRef]
16. Peng, Y.; Le, Z.; Wen, M.; Zhang, D.; Chen, Z.; Bin Wu, H.; Li, H.; Lu, Y. Mesoporous single-crystal-like TiO_2 mesocages threaded with carbon nanotubes for high-performance electrochemical energy storage. *Nano Energy* **2017**, *35*, 44–51. [CrossRef]
17. Lei, C.; Chen, Z.; Sohn, H.; Wang, X.; Le, Z.; Weng, D.; Shen, M.; Wang, G.; Lu, Y. Better lithium-ion storage materials made through hierarchical assemblies of active nanorods and nanocrystals. *J. Mater. Chem. A* **2014**, *2*, 17536–17544. [CrossRef]
18. Chen, H.; Jiang, J.; Zhang, L.; Xia, D.; Zhao, Y.; Guo, D.; Qi, T.; Wan, H. In situ growth of $NiCo_2O_4$ nanotube arrays on Ni foam for supercapacitors: Maximizing utilization efficiency at high mass loading to achieve ultrahigh areal pseudocapacitance. *J. Power Sources* **2014**, *254*, 249–257. [CrossRef]
19. Zou, R.; Xu, K.; Wang, T.; He, G.; Liu, Q.; Liu, X.; Zhang, Z.; Hu, J. Chain-like $NiCo_2O_4$ nanowires with different exposed reactive planes for high-performance supercapacitors. *J. Mater. Chem. A* **2013**, *1*, 8560. [CrossRef]

20. Zhang, Y.; Ma, M.; Yang, J.; Su, H.; Huang, W.; Dong, X. Selective synthesis of hierarchical mesoporous spinal $NiCo_2O_4$ for high-performance supercapacitors. *Nanoscale* **2014**, *6*, 4303–4308. [CrossRef]

21. Lei, Y.; Li, J.; Wang, Y.; Gu, L.; Chang, Y.; Yuan, H.; Xiao, D. Rapid Microwave-Assisted Green Synthesis of 3D Hierarchical Flower-Shaped $NiCo_2O_4$ Microsphere for High-Performance Supercapacitor. *ACS Appl. Mater. Interfaces* **2014**, *6*, 1773–1780. [CrossRef] [PubMed]

22. Yedluri, A.K.; Kim, H.-J. Enhanced electrochemical performance of nanaoplate nickel cobaltite ($NiCo_2O_4$) supercapacitor applications. *RSC Adv.* **2019**, *9*, 1115–1122. [CrossRef]

23. Yuan, C.; Li, J.; Hou, L.; Zhang, X.; Shen, L.; Lou, X.W. Ultrathin Mesoporous $NiCo_2O_4$ Nanosheets Supported on Ni Foam as Advanced Electrodes for Supercapacitors. *Adv. Funct. Mater.* **2012**, *22*, 4592–4597. [CrossRef]

24. Zhu, Y.; Ji, X.; Wu, Z.; Song, W.; Hou, H.; Wu, Z.; He, X.; Chen, Q.; Banks, C.E. Spinal $NiCo_2O_4$ for use as a high-performance supercapacitor electrode material: Understanding of its electrochemical properties. *J. Power Sources* **2014**, *267*, 888–900. [CrossRef]

25. Liu, R.; Ma, L.; Huang, S.; Mei, J.; Li, E.; Yuan, G. Large areal mass and high scalable and flexible cobalt Oxide/Graphene/Bacterial cellulose electrode for supercapacitors. *J. Phys. Chem. C* **2016**, *120*, 28480–28488. [CrossRef]

26. Cui, C.; Xu, J.; Wang, L.; Guo, D.; Mao, M.; Ma, J.; Wang, T. Growth of $NiCo_2O_4$@$MnMoO_4$ Nanocolumn Arrays with Superior Pseudocapacitor Properties. *ACS Appl. Mater. Interfaces* **2016**, *8*, 8568–8575. [CrossRef] [PubMed]

27. Mai, L.Q.; Yang, F.; Zhao, Y.L.; Xu, X.; Xu, L.; Luo, Y.Z. Hierarchical $MnMoO_{(4)}$/$CoMoO_{(4)}$ heterostructured nanowires with enhanced supercapacitor performance. *Nat. Commun.* **2011**, *2*, 381–385. [CrossRef] [PubMed]

28. Yuan, Y.; Wang, W.; Yang, J.; Tang, H.; Ye, Z.; Zeng, Y.; Lu, J. Three-Dimensional $NiCo_2O_4$@$MnMoO_4$ Core–Shell Nanoarrays for High-Performance Asymmetric Supercapacitors. *Langmuir* **2017**, *33*, 10446–10454. [CrossRef]

29. Cheng, J.; Lu, Y.; Qiu, K.; Yan, H.; Xu, J.; Han, L.; Liu, X.; Luo, J.; Kim, J.-K.; Luo, Y. Hierarchical Core/Shell $NiCo_2O_4$@$NiCo_2O_4$ Nanocactus Arrays with Dual-functionalities for High Performance Supercapacitors and Li-ion Batteries. *Sci. Rep.* **2015**, *5*, 12099. [CrossRef]

30. Gu, Z.; Zhang, X. $NiCo_2O_4$@$MnMoO_4$ core–shell flowers for high performance supercapacitors. *J. Mater. Chem. A* **2016**, *4*, 8249–8254. [CrossRef]

31. Zhang, Z.; Liu, Y.; Huang, Z.; Ren, L.; Qi, X.; Wei, X.; Zhong, J. Facile hydrothermal synthesis of $NiMoO_4$@$CoMoO_4$ hierarchical nanospheres for supercapacitor applications. *Phys. Chem. Chem. Phys.* **2015**, *17*, 20795–20804. [CrossRef] [PubMed]

32. Li, W.; Yang, F.; Hu, Z.; Liu, Y. Template synthesis of C@$NiCo_2O_4$ hollow microsphere as electrode material for supercapacitor. *J. Alloys Compd.* **2018**, *749*, 305–312. [CrossRef]

33. Karmakar, S.; Varma, S.; Behera, D.; Varma, S. Investigation of structural and electrical transport properties of nano-flower shaped $NiCo_2O_4$ supercapacitor electrode materials. *J. Alloys Compd.* **2018**, *757*, 49–59. [CrossRef]

34. Yuan, C.; Li, J.; Hou, L.; Yang, L.; Shen, L.; Zhang, X. Facile template-free synthesis of ultralayered mesoporous nickel cobaltite nanowires towards high-performance electrochemical capacitors. *J. Mater. Chem.* **2012**, *22*, 16084–16090. [CrossRef]

35. Liu, X.; Liu, J.; Sun, X. $NiCo_2O_4$@NiO hybrid arrays with improved electrochemical performance for pseudocapacitors. *J. Mater. Chem. A* **2015**, *3*, 13900–13905. [CrossRef]

36. Li, J.; Xiong, S.; Liu, Y.; Ju, Z.; Qian, Y. High Electrochemical Performance of Monodisperse $NiCo_2O_4$ Mesoporous Microspheres as an Anode Material for Li-Ion Batteries. *ACS Appl. Mater. Interfaces* **2013**, *5*, 981–988. [CrossRef] [PubMed]

37. Choudhury, T.; Saied, S.O.; Abbot, A.M.; Sullivan, J.L. Reduction of oxides of iron, cobalt, titanium and niobium by low-energy ion bombardment. *J. Phys. D Appl. Phys.* **1989**, *22*, 1185–1195. [CrossRef]

38. Li, X.; Wei, J.; Li, Q.; Zheng, S.; Xu, Y.; Du, P.; Chen, C.; Zhao, J.; Xue, H.; Xu, Q.; et al. Nitrogen-Doped Cobalt Oxide Nanostructures Derived from Cobalt-Alanine Complexes for High-Performance Oxygen Evolution Reactions. *Adv. Funct. Mater.* **2018**, *28*, 1800886. [CrossRef]

Article

Hybrid Model Predictive Control Strategy of Supercapacitor Energy Storage System Based on Double Active Bridge

Lujun Wang [1,*], Jiong Guo [1], Chen Xu [1], Tiezhou Wu [1] and Huipin Lin [2]

[1] Hubei Key Laboratory for High-Efficiency Utilization of Solar Energy and Operation Control of Energy Storage System, Hubei University of Technology, Wuhan 430068, China; winddreamleaf@163.com (J.G.); xuchen961215@163.com (C.X.); wtz315@163.com (T.W.)

[2] College of Electrical Engineering, Zhejiang University, Hangzhou 310027, China; linhuipin@zju.edu.cn

* Correspondence: wanglujun@zju.edu.cn; Tel.: +86-153-2731-2005

Received: 19 April 2019; Accepted: 30 May 2019; Published: 4 June 2019

Abstract: In order to solve the problem of which the dynamic response of a supercapacitor (SC) is limited due to the mismatch dynamic characteristics between the DC/DC converter and supercapacitor in an energy storage system, this paper proposes a hybrid model predictive control strategy based on a dual active bridge (DAB). The hybrid model predictive control model considers the supercapacitor and DAB in a unified way, including the equivalent series resistance and capacitance parameters of the SC. The method can obtain a large charging and discharging current of the SC, thereby not only improving the overall response speed of the system, but also expanding the actual capacity utilization range of the SC. The simulation results show that compared with the model prediction method of the dual active bridge converter, the proposed control method can effectively improve the overall response speed of the system, which can be improved by at least 0.4 ms. In addition, the proposed method increases the actual upper limit of the SC voltage, reduces the actual lower limit of the SC voltage, and then expands the actual capacity utilization range of the SC by 18.63%. The proposed method has good application prospects in improving the dynamic response performance of energy storage systems.

Keywords: dual active bridge; SC; energy storage; fast response; capacity utilization

1. Introduction

Supercapacitors (SCs) have developed rapidly in recent years due to their large capacity, high power density [1], high charge-discharge rate, high conversion efficiency, wide operating temperature range, no pollution, and convenient control. However, due to the wide range of terminal voltages during the storage and release of energy in an SC, the systems used in actual energy storage systems are usually equipped with bidirectional DC/DC converters to implement power conversion [2].

In addition, since an SC voltage unit is low [3], many units should be connected in series in actual use, causing the problem of equalization. For some applications, where the DC bus voltage is relatively large, such as a three-phase grid-connected inverter, a high-gain bidirectional DC converter suitable for a wide voltage range is needed to adjust the output voltage to meet the energy requirements of the DC bus [4]. In recent years, dual active bridge bidirectional DC-DC converters are commonly used in medium power applications [5], which enable a soft switching operation over a wide voltage range. Additionally, the transformer turns ratio can be used to obtain a high gain output [6]. In addition, the DAB converter has the advantages of a high power-density, bidirectional transmission power, high efficiency, and structural symmetry [7,8]. Due to the large number of switches in a DAB converter, the control process is generally complicated, so the response time of the converter is longer than with an

SC [9], which does not reflect the fast response characteristics of SCs. In order to improve the dynamic response speed of the converter when the load is abrupt or the input voltage fluctuates, some scholars have proposed related control methods.

In [10], an inductor current boundary control method was proposed based on tradition inductor current peak control and single-phase shift control. The boundary control method significantly speeds up the dynamic response of the DC-DC converter when the load suddenly changes, but the method is too complicated and requires up to five Hall sensors. In [11], a load current feedforward control method was proposed based on direct power control, which introduces the load current into the control system, significantly improving the load response speed of the converter. However, it did not analyze the response characteristics of the converter when the input voltage was changed. In [12], based on the idea of direct power control, a virtual direct power control strategy was proposed, which introduces input voltage and load current information at the same time, so that the converter achieves a fast dynamic response. In [13], based on the working principle of a dual-active bridge DC-DC converter, the principle of single-phase shift control, and the steady-state characteristics of the output current, an output current control method based on the input voltage feedforward was proposed to improve the dynamic characteristics of the converter. This method is based on the research of the dual active bridge DC-DC converter as the current source. In [14], a model predictive control strategy was proposed based on the establishment of the state space average model of a DAB by analyzing the characteristics of a DAB and using the traditional phase shift control as an example. The proposed method significantly improved the dynamic characteristics of the converter for the input voltage's abrupt change and the load's abrupt change compared with the traditional voltage closed-loop control method.

In order to improve the overall response speed of a super-capacitor energy storage system based on dual active bridges, a hybrid model predictive control strategy with a fast dynamic response time is proposed. First, the overall structure model of the energy storage system based on a combined appropriate SC equivalent model and a DAB converter model is established in this paper. Then, the working principle of a DAB based on traditional phase shift control is analyzed, and the state space average model of the energy storage system composed of a DAB converter and SC is established. By selecting the appropriate evaluation target and establishing the cost function, a model predictive control strategy incorporating the parameters of the SC equivalent model is proposed to manage the energy of the system. Finally, the effectiveness of the proposed control strategy is verified by simulations in Matlab (9.2.0.538062 (R2017a), MathWorks, Natick, Massachusetts, America) software and experiments.

2. Working Principle of the DAB-SC System

2.1. SC Model

The structure of the SC unit is mainly composed of electrodes, electrolytes, diaphragms, and packaging, and each component has an important impact on its performance. Thus, it is necessary to build a precise and applicable SC model in circuit analysis. Scholars have proposed many SC equivalent models based on the physical structure and charging and discharging conditions, such as the first-order classical resistor-capacitor (RC) model, RC transmission line model, multi-branch RC parallel model, frequency domain model [15], and artificial neural network (ANN) model. By combining various SC models to obtain a general model, the general SC equivalent circuit model can be simplified based on characteristics, such as: Large-scale modularization of SCs, high charging and discharging power, and high charging and discharging frequency. The simplified model is shown in Figure 1.

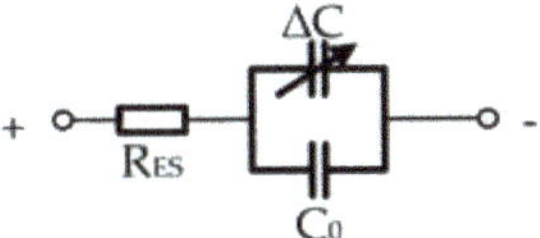

Figure 1. Simplified capacitor circuit model of the SC unit.

The use of SCs for large-scale energy storage requires identification of the SC parameters. In the simplified equivalent circuit model of the SC unit, R_{ES} is the equivalent series resistance, which not only characterizes the heat loss inside an SC, but also causes different voltage drops as the discharge current changes. Thus, R_{ES} affects the maximum value of the discharge current to a certain extent. The magnitude of R_{ES} generally does not change with external factors [16], and is usually measured by a classical model.

When the charge and discharge conditions of the SC are determined, the main factor that affects the value of the variable capacitor in the equivalent circuit is the actual operating voltage [17]. In order to identify the parameters of the SC unit, several sets of tests are needed, by charging the SC unit to different voltage levels with a constant frequency.

According to Equation (1), the actual capacitance of an SC unit in different sets of tests can be calculated:

$$C_{SC} = \frac{Q_{SC}}{\Delta U_{SC}} = \frac{\int I_{SC}dt}{\Delta U_{SC}} = \frac{I_{SC}\Delta T}{\Delta U_C - \Delta U_{ESR}}. \tag{1}$$

In Equation (1), Q_{SC} is the amount of charge of an SC; ΔU_{SC} is the actual voltage change; I_{SC} is the SC current; ΔT is the charging and discharging time; ΔU_C is the total voltage change; ΔU_{ESR} is the voltage drop caused by E_{SR}.

After processing the (U_{charge}, C_{actual}) data-points by the linear fitting method, the relationship between the capacitance and voltage of a SC can be obtained as shown in Equation (2):

$$C_{actual} = \Delta C + C_0 = K_v U_{charge} + C_0. \tag{2}$$

In Equation (2), C_{actual} is the actual capacitance of an SC at different voltages; ΔC is the variable capacitance; C_0 is the initial capacitance of an SC; K_v is the coefficient of variation of the SC capacitance with the voltage; U_{charge} is the SC's actual voltage.

In this paper, through the linear fitting of graphic data in [18], the voltage and capacitance data of Maxwell's 1 F SC were obtained, and then the data of the 125 V, 63 F SC, which was used in this paper, was obtained according to the string and parallel relationship of the capacitor. The 125 V, 63 F SC is made up of 46 modules in series, and the module was obtained by paralleling 2900 Maxwell's 1 F, 2.7 V SC units. Additionally, the obtained data of the 125 V, 63 F SC were linearly fitted to obtain a relation as shown in Equation (3):

$$C_{actual} = 0.1127U_{charge} + 81.384. \tag{3}$$

2.2. Principle Analysis of DAB

The DAB converter acts as an interface between the SC and DC bus. Generally, there are two basic control methods for the DAB converter: The first is pulse width modulation control; the second is phase shift control. Phase shift control is simpler and easier to implement than the first. Therefore, most DAB controls adopt the phase shift control method [19]. Taking single-phase shift as an example, when a single DAB is in normal operation, the switching frequencies of the H-bridges on both sides of the isolation transformer are the same, and the two H-bridges output square-wave voltages, u_{ab} and u_{cd}, respectively, with a duty ratio of 50%. When the phase of the u_{ab} leads the u_{cd}, the energy is forwardly transmitted; that is, from u_{dc} to u_o. Otherwise, the energy is reversely transmitted. Therefore, by controlling the degree of the phase shift angle between the two square wave voltages, the transferred energy can be controlled.

Figure 2 shows the topology of the DAB-SC system. Figure 3 shows the main working waveform of the converter with the phase shift control method. The two switches on each bridge arm are complementarily turned on, and the diagonal switches are turned on or off at the same time. The power is controlled by controlling the phase shift angle between the midpoint voltages, u_{ab} and u_{cd}, of the primary and secondary side arms. T_s is half of the switching frequency, $T_s = 1/(2f_s)$, and D represents the phase shift between u_{ab} and u_{cd}, where $D = t_{on}/T_s$, and $-1 < D < 1$. The working principles of the converter when transmitting power in the forward and reverse directions are similar. This paper utilized the forward direction as an example to analyze the operating characteristics of the DAB converter. Before the analysis, the following assumptions were made: (1) The converter is in steady state operation; (2) all switches are regarded as ideal switches with a reverse parallel diode, and the parasitic capacitance is connected in parallel; (3) the size of L is equal to the sum of the leakage inductance of the transformer and the applied inductance; (4) the transformer is ideal, the excitation current is zero; and (5) $u_{dc} > nu_o$, where n is the turns ratio of the transformer.

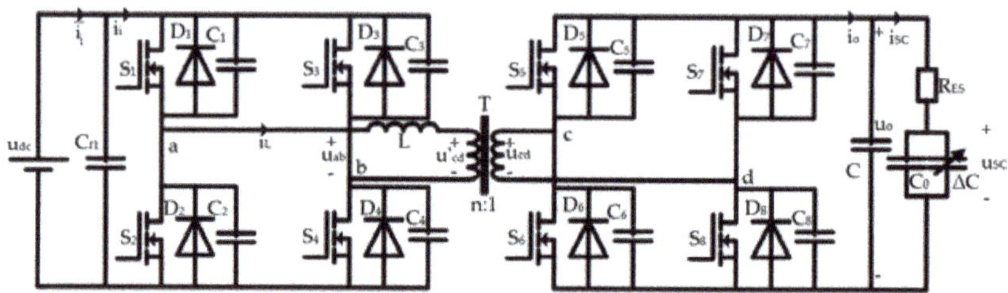

Figure 2. Topology of the DAB-SC system.

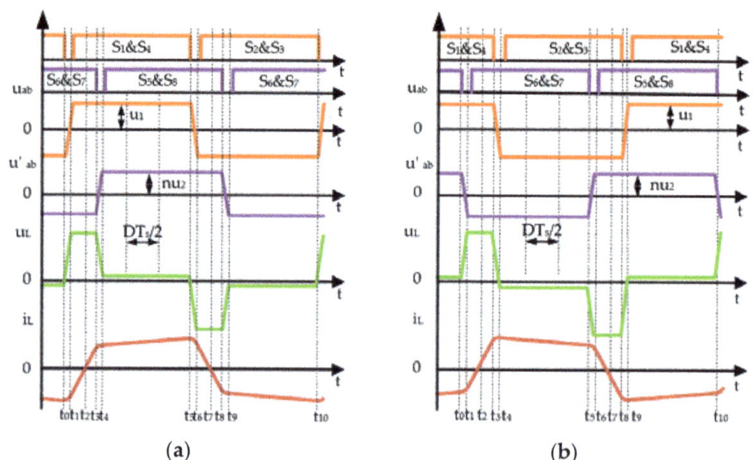

(a)

(b)

Figure 3. The main working waveform of the DAB converter with phase shift control: (**a**) Forward power transfer; (**b**) reverse power transfer.

When the converter is working in the forward direction, the driving signals of S_1, S_4 and S_2, S_3 are ahead of S_5, S_8 and S_6, S_7 respectively, as shown in Figure 3a. In the figure, u'_{cd} is the voltage of u_{cd} converted to the primary side; u_L is the voltage across the inductor, L; and i_L is the inductor current.

Under steady-state conditions, the relationship between the transmission power of the DAB and the amount of phase shift is as shown in Equation (4):

$$P_o = \frac{nu_{dc}u_o}{2f_sL}D(1-|D|).$$

(4)

It can be known from Equation (4) that the transmission power reaches maximum when $D = 0.5$ in the forward transmission, and when $D = -0.5$ in the reverse transmission. Typically, the DAB converter operates in the range of $D \in [-0.5, 0.5]$.

The output power of the DAB can also be expressed as $P_o = u_o i_o$, and the relationship between the output current of the DAB converter and the phase shift amount can be obtained as shown in Equation (5):

$$i_o = \frac{n u_{dc}}{2 f_s L} D(1 - |D|). \tag{5}$$

In the DAB converter, there are two key parameters, which are the transformer ratio, n, and the buffer inductance, L. When the input voltage of the DAB converter is equal to the value of the output voltage converted to the primary side, the current stress of the device is minimized. By considering the actual fluctuation of the input and output voltages, the transformer ratio can be selected to make the rated input voltage equal to the value of the rated output voltage converted to the primary side. For example, when the rated voltages of the input and output voltages are 700 V and 75 V, respectively, the transformer ratio is determined to be n = 700/75 ≈ 9.

The zero voltage switch (ZVS) range of the device and the suppression of the reactive power are mutually contradictory. A larger ZVS range of a device requires larger inductor values; however, for the suppression of the reactive power, the smaller the inductor value, the better [20], which is a contradiction. The focus of this paper was to study the response speed of the system, and thus the influence of ZVS was not considered. Therefore, a small inductance value as selected for the research.

2.3. State Space Average Model of DAB-SC

In order to establish the state space average model of the DAB, the output capacitor voltage and the inductor current were selected for modeling. It can be seen from Figure 3a that when the voltage, u_{ab}, leads the voltage, u_{cd}, the energy is transmitted forward, $0 < D < 0.5$. The overall DAB-SC system has six working states in one switching cycle, and a system of differential equations can be established in each working state. According to Kirchhoff's law, the differential equations, Equations (6)–(11), can be obtained:

$$\begin{cases} \frac{di_L}{dt} = \frac{n u_o}{L} + \frac{u_{dc}}{L} \\ \frac{du_o}{dt} = -\frac{n i_L}{C} - \frac{u_o - u_{SC}}{R_{ES}C} \end{cases} \quad t \in [0, \frac{D}{2} T_s], \tag{6}$$

$$\begin{cases} \frac{di_L}{dt} = \frac{n u_o}{L} + \frac{u_{dc}}{L} \\ \frac{du_o}{dt} = -(\frac{n i_L}{C} - \frac{u_{SC} - u_o}{R_{ES}C}) \end{cases} \quad t \in [\frac{D}{2} T_s, D T_s], \tag{7}$$

$$\begin{cases} \frac{di_L}{dt} = -\frac{n u_o}{L} + \frac{u_{dc}}{L} \\ \frac{du_o}{dt} = \frac{n i_L}{C} - \frac{u_o - u_{SC}}{R_{ES}C} \end{cases} \quad t \in [D T_s, T_s], \tag{8}$$

$$\begin{cases} \frac{di_L}{dt} = -\frac{n u_o}{L} - \frac{u_{dc}}{L} \\ \frac{du_o}{dt} = \frac{n i_L}{C} - \frac{u_o - u_{SC}}{R_{ES}C} \end{cases} \quad t \in [T_s, (1 + \frac{D}{2}) T_s], \tag{9}$$

$$\begin{cases} \frac{di_L}{dt} = -\frac{n u_o}{L} - \frac{u_{dc}}{L} \\ \frac{du_o}{dt} = \frac{n i_L}{C} + \frac{u_{SC} - u_o}{R_{ES}C} \end{cases} \quad t \in [(1 + \frac{D}{2}) T_s, (1 + D) T_s], \tag{10}$$

$$\begin{cases} \frac{di_L}{dt} = \frac{n u_o}{L} - \frac{u_{dc}}{L} \\ \frac{du_o}{dt} = -\frac{n i_L}{C} - \frac{u_o - u_{SC}}{R_{ES}C} \end{cases} \quad t \in [(1 + D) T_s, 2 T_s]. \tag{11}$$

For each differential equation, it can only represent the voltage–current relationship in this operating state. Therefore, it is necessary to establish a differential equation that can describe the characteristics of the converter over the entire switching period. It is worth noting that in the DAB, the inductor current is a pure alternative current (AC) component whose average value is zero during one switching cycle, so the average value of the inductor current is meaningless. Therefore, a simplified reduction model of the DAB module can be obtained by eliminating the inductor current as shown in Equation (12):

$$\frac{d\langle u_o \rangle}{dt} = \frac{nD(1-D)\langle u_{dc} \rangle}{2Lf_sC} - \frac{\langle u_o \rangle - \langle u_{SC} \rangle}{R_{ES}C}.$$

(12)

Similarly, when the voltage, u_{cd}, leads u_{ab}, the energy is reversely transmitted; that is, $-0.5 < D < 0$, and the simplified reduced-order model of the DAB module is shown in Equation (13):

$$\frac{d\langle u_o \rangle}{dt} = \frac{nD(1-D)\langle u_{dc} \rangle}{2Lf_sC} - \frac{\langle u_o \rangle - \langle u_{SC} \rangle}{R_{ES}C}.$$

(13)

In conclusion, when $-0.5 < D < 0.5$, the simplified reduced-order model of the DAB module is as shown in Equation (14):

$$\frac{d\langle u_o \rangle}{dt} = \frac{nD(1-|D|)\langle u_{dc} \rangle}{2Lf_sC} - \frac{\langle u_o \rangle - \langle u_{SC} \rangle}{R_{ES}C}.$$

(14)

3. Management Strategy

The state of charge (SOC) reflects the remaining capacity of the SC. In a photovoltaic system equipped with energy storage, by accurately estimating the SOC, how long the SC can work under a current load can be determined. There is a definite relationship between the energy, E_{SC}, stored by the SC and the terminal voltage, as shown in Equation (15):

$$E_{SC} = \frac{C_{SC}U_o^2}{2}.$$

(15)

Therefore, it is only necessary to detect the SC voltage to accurately determine its SOC. The relationship between the voltage of the SC and the SOC is as shown in Equation (16):

$$SOC = \frac{\frac{C_{SC}U_o^2}{2}}{\frac{C_{SC}U_N^2}{2}} = \left(\frac{U_o}{U_N}\right)^2.$$

(16)

In Equation (16), C_{SC} is the capacity of the SC, U_o is the SC's terminal voltage, and U_N is the rated voltage of the SC.

Based on the relationship between the terminal voltage and the SOC, the system can be managed by controlling the terminal voltage. Six voltage limits were set: The DC bus voltage upper limit, U_{dc_max}; lower limit, U_{dc_min}; SC upper limit voltage, U_{sc_max}; lower limit voltage, U_{sc_min}; upper limit warning voltage, U_{sc_h}; and lower limit warning voltage, U_{sc_l}. The working principle is as follows: When the DC bus voltage exceeds U_{dc_max} and the SC voltage is close to U_{sc_min}, the SC can be charged with a large current. As the SC voltage increases, the phase shift duty ratio of the DAB converter is changed by predictive control in order to prevent overcharging and to increase the upper limit of the actual SC charging voltage; when the DC bus voltage exceeds U_{dc_min}, and the SC voltage is close to U_{sc_max}, the SC performs a large current discharge. As the SC voltage decreases, the phase shift duty ratio of the DAB converter is also changed by predictive control in order to prevent over discharging and to decrease the lower limit of the actual SC charging voltage.

According to the State Grid Corporation "DC Power System Technical Supervision Regulations" and other relevant standards and specifications, when the DC bus power supply is continuously supplied, normal fluctuation of the grid voltage should be no more than 10% of the rated voltage [21]. The rated DC bus voltage of the microgrid in this paper was set at 700 V with a fluctuation of ±5%; that is, U_{dc_min} = 665 V, and U_{dc_max} = 735 V. The SC had a rated voltage of 125 V and the upper limit warning voltage and the lower limit warning voltage were 40%–80%.

In summary, the coordinated control strategy of the energy storage systems is as follows: a) In the shutdown state of an energy storage system, the DAB converter is in an idle state and is not charged or discharged; b) in the SC charging area, when the DC bus voltage, U_{dc}, satisfied that U_{dc} > 735 V, and $SOCsc \leq 80\%$, the SC is charged to absorb excess power in the DC bus; and c) in the SC discharge area, when the DC bus voltage satisfied that U_{dc} < 665 V, and SOC_{sc} > 40%, the SC discharges, supplementing the power shortage in the DC bus.

4. Model Predictive Control

Model predictive control (MPC) is an active control strategy based on the mathematical model of the converter and is adjusted by predicting the change trend of the control variables. It predicts the optimality of the next moment based on the circuit parameters and the sampling information at the current time to minimize the deviation between the output and the reference. There are many forms of the MPC algorithm, and all MPC algorithms have three basic characteristics: Prediction model, rolling optimization, and feedback correction. A schematic diagram of the MPC strategy of the DAB-SC system studied in this paper is shown in Figure 4, where u_o is the DAB output voltage, u_{dc} is the DC bus voltage, and i_{SC} is the SC current.

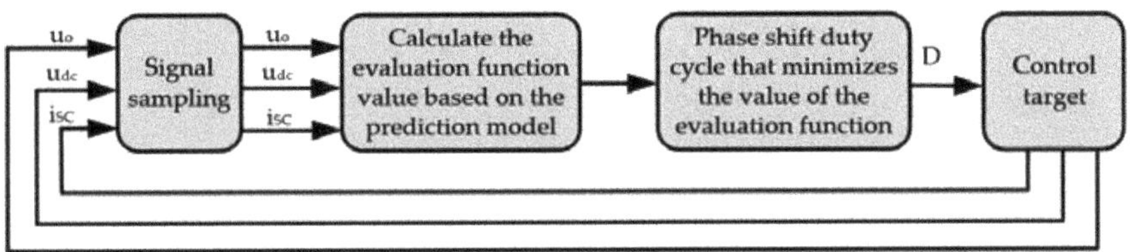

Figure 4. Schematic diagram of the MPC strategy.

4.1. Establishment of the Prediction Model for the DAB-SC System

The derivative of the output voltage in Equation (14) reflects the trend of the output voltage. Additionally, the Euler forward method can be used to discretize Equation (14), and then Equation (17) can be obtained:

$$\frac{du_o}{dt} = \frac{u_o(t_{k+1}) - u_o(t_k)}{t_{k+1} - t_k} = \frac{u_o(t_{k+1}) - u_o(t_k)}{2T_s}. \tag{17}$$

Equation (17) can be simplified with Equation (14) to obtain Equation (18):

$$u_o(t_{k+1}) = \frac{nD(1 - |D|)u_{dc}(t_k)}{2LCf_s^2} - \frac{u_o(t_k) - u_{SC}(t_k)}{R_{ES}Cf_s} + u_o(t_k). \tag{18}$$

The SC has a high-power density characteristic and can provide a large instantaneous current, thus the current flowing through SC at time t_{k+1} was set to be $i_{SC}(t_{k+1})$. Then, Equation (19) can be established:

$$\frac{u_o(t_k) - u_{SC}(t_k)}{R_{ES}} = i_{SC}(t_{k+1}). \tag{19}$$

By substituting Equation (19) into Equation (18), Equation (20) can be obtained:

$$u_o(t_{k+1}) = \frac{nD(1 - |D|)u_{dc}(t_k)}{2LCf_s^2} - \frac{i_{SC}(t_{k+1})}{Cf_s} + u_o(t_k). \tag{20}$$

In Equation (20), $u_o(t_k)$ and $u_{dc}(t_k)$ represent the DAB output voltage and DC bus voltage, respectively, sampled at time t_k; $u_o(t_{k+1})$ represents the predicted value of the SC voltage at time t_{k+1}.

According to the general characteristics of the capacitor, the equation of the equivalent capacitor in the SC equivalent model can be obtained, as shown in Equation (21):

$$C_{SC}\frac{du_{SC}}{dt} = i_{SC}, \tag{21}$$

where dt = 2Ts in Equation (21). In order to reflect the fast response characteristic of the SC, we chose i_{SC} in Equation (21) to be $i_{SC}(t_{k+1})$. Then, Equation (22) can be obtained:

$$[u_{SC}(t_{k+1}) - u_{SC}(t_k)]C_{SC} = 2T_s i_{SC}(t_{k+1}) = \frac{i_{SC}(t_{k+1})}{f_s}. \tag{22}$$

According to Equation (2), the relationship between the capacitance of the SC and the actual voltage across the SC is as shown in Equation (23):

$$C_{SC} = K_v u_o(t_k) + C_0. \tag{23}$$

According to Equation (22) and Equation (23), taking the average value of each state in the case of the charging and discharging of the SC, Equation (24) can be obtained:

$$[u_o(t_{k+1}) - i_{SC}(t_{k+1})R_{ES} - u_o(t_k) + i_{SC}(t_k)R_{ES}]\cdot[K_v u_o(t_k) + C_0] = \frac{i_{SC}(t_{k+1})}{f_s}. \tag{24}$$

By joining Equation (20) and Equation (24) together to eliminate $i_{SC}(t_k)$, Equation (25) can be obtained:

$$u_o(t_{k+1}) = \frac{nD(1 - |D|)u_{dc}(t_k)}{2LCf_s^2}\left(1 - \frac{1}{A}\right) - \frac{1}{A}i_{SC}(t_k)R_{ES} + u_o(t_k), \tag{25}$$

where the value of A is as shown in Equation (26):

$$A = \frac{C}{K_v u_o(t_k) + C_0} + Cf_s R_{ES} + 1. \tag{26}$$

4.2. Phase Shift Optimization Calculation and Analysis

For energy storage systems with a DAB, the primary control objective is to stabilize the DC bus voltage. When the bus voltage is too high, the SC needs to be charged. When the bus voltage is low, the SC is required to discharge.

Therefore, different SC reference voltages were selected in this paper. When the bus voltage is high, the SC reference value takes the upper limit warning voltage, U_{sc_h}; when the bus voltage is low, the SC reference value takes the lower limit warning voltage, U_{sc_l}. The value of the SC reference value is shown in Equation (27):

$$U_o^* = \begin{cases} U_{SC\_h}, u_{dc}(k) > U_{dc\_max}; \\ U_{SC\_l}, u_{dc}(k) < U_{dc\_min}. \end{cases} \tag{27}$$

The deformation of Equation (27) is given by Equation (28):

$$U_o^* = \frac{U_{SC\_h} + U_{SC\_l}}{2} + \frac{U_{SC\_h} - U_{SC\_l}}{2}\cdot B. \tag{28}$$

The value of B in Equation (28) is as shown in Equation (29):

$$B = \frac{\frac{u_{dc}(k) - U_{dc\_max}}{|u_{dc}(k) - U_{dc\_max}|} + \frac{u_{dc}(k) - U_{dc\_min}}{|u_{dc}(k) - U_{dc\_min}|}}{2} = \begin{cases} 1, u_{dc}(k) > U_{dc\_max}; \\ -1, u_{dc}(k) < U_{dc\_min}. \end{cases} \tag{29}$$

Therefore, a cost function is established based on the prediction model of Equation (26) of the output voltage as shown in Equation (30):

$$J(k) = (u_o(t_{k+1}) - U_o^*)^2. \tag{30}$$

In order to minimize the deviation between the output voltage and the reference voltage in the next moment, the cost function needs to be minimized; that is, the selected optimal phase shift amount should minimize the cost function value. By combining Equations (25), (28), and (29) and deriving Equation (30), the value of D can be obtained as in Equation (31):

$$D = \text{B} \cdot \frac{1 - \sqrt{1 - 4BM}}{2}. \tag{31}$$

The value of M in Equation (31) is as shown in Equation (32):

$$M = \frac{A}{A - 1} \cdot \frac{2LCf_s^2}{nu_{dc}(k)} [U_o^* - u_o(t_k) + \frac{1}{A} \cdot i_{SC}(t_k) R_{ES}]. \tag{32}$$

A block diagram of the MPC structure constructed according to Equation (32) is shown in Figure 5.

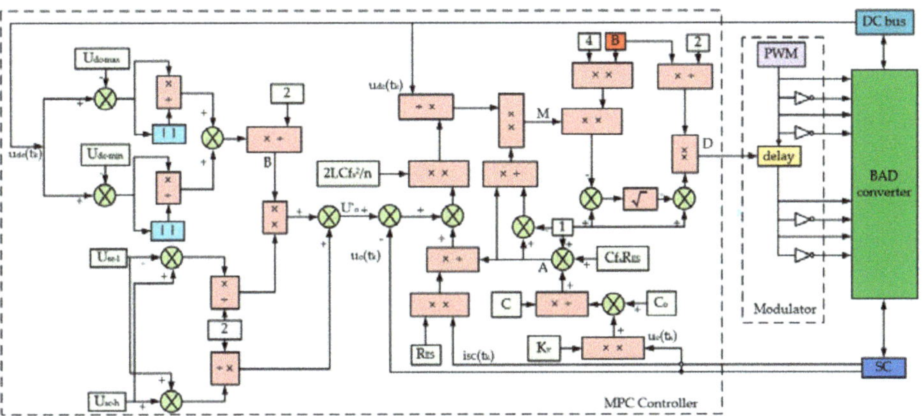

Figure 5. MPC structure block diagram.

It can be seen from the above analysis that the optimal phase shift of the DAB converter in the proposed method is obtained based on the sampling voltage, sampling current, and circuit parameters of the DAB-SC system. Additionally, the parameters of the SC can reflect the fast response characteristics of the SC. Compared with predictive control based on the simple construction of a DAB converter, the proposed control strategy makes the overall response of the DAB-SC system faster.

5. Simulation Analysis

A simulation model of the DAB-SC system, as shown in Figure 6, was built in MATLAB/simulink according to the topology diagram of Figure 2 and the control block diagram of Figure 5. The parameters of the simulation are shown in Table 1. Additionally, the experimental simulation pulse signal generator had a frequency of 20 kHz and a duty cycle of 50%.

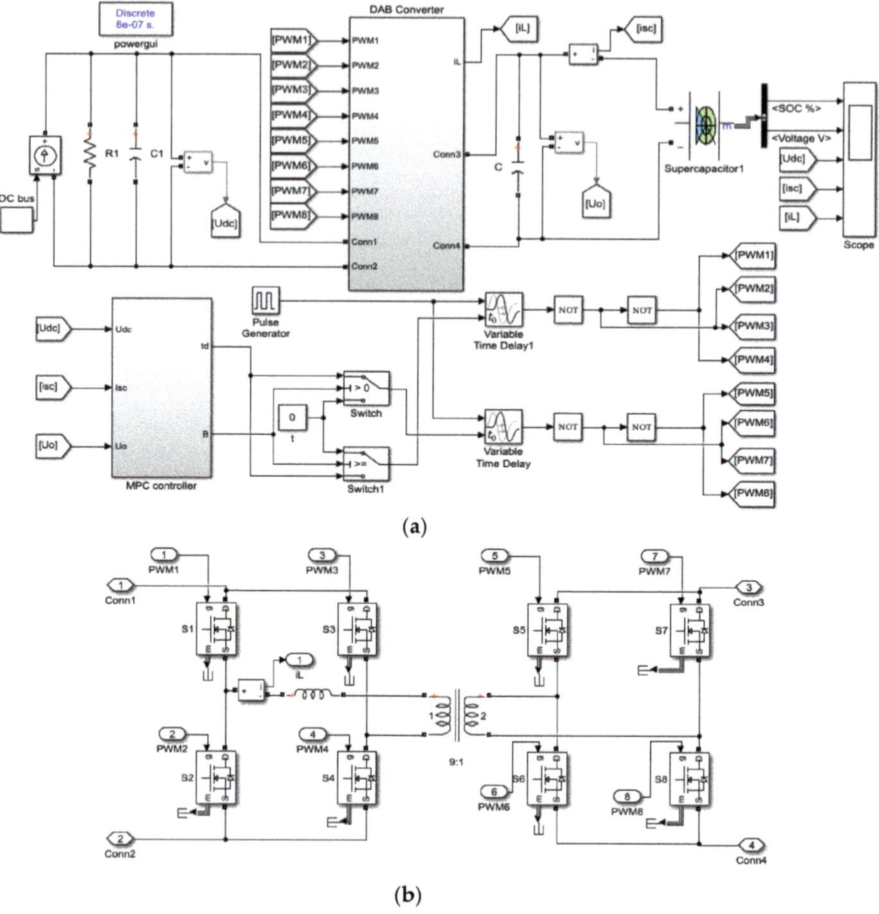

Figure 6. Simulation model of the DAB-SC system based on Matlab/Simulink: (**a**) The overall model; (**b**) model of the DAB converter.

Table 1. Simulation Parameters.

Symbol	Quantity
f_s	20 kHz
U_{dc}	700 V
C	3×10^{-3} F
L	6.815×10^{-6} H
n	9
R_{ES}	$0.018\ \Omega$
K_v	0.1127
C_0	81.384

5.1. SC Discharge in the Normal Working Range

When the DC bus voltage, U_{dc}, is lower than the lower limit voltage of 665 V, which is 651.7 V, and the initial value of the voltage, U_o, on the SC side is 75 V, the dynamic response time of the system is 1.9 ms when the bus voltage is stabilized to 665 V with only a prediction model of the DAB converter. The simulation waveform is shown in Figure 7a. Under the same conditions, using the overall prediction model composed of the DAB converter and the SC to control the SC discharge to stabilize the bus voltage to 665 V, the dynamic response time of the system is 1.5 ms, and the simulation waveform is shown in Figure 7b. Under these simulation conditions, a detailed waveform of the drive signal is shown in Figure A1 (see Appendix A).

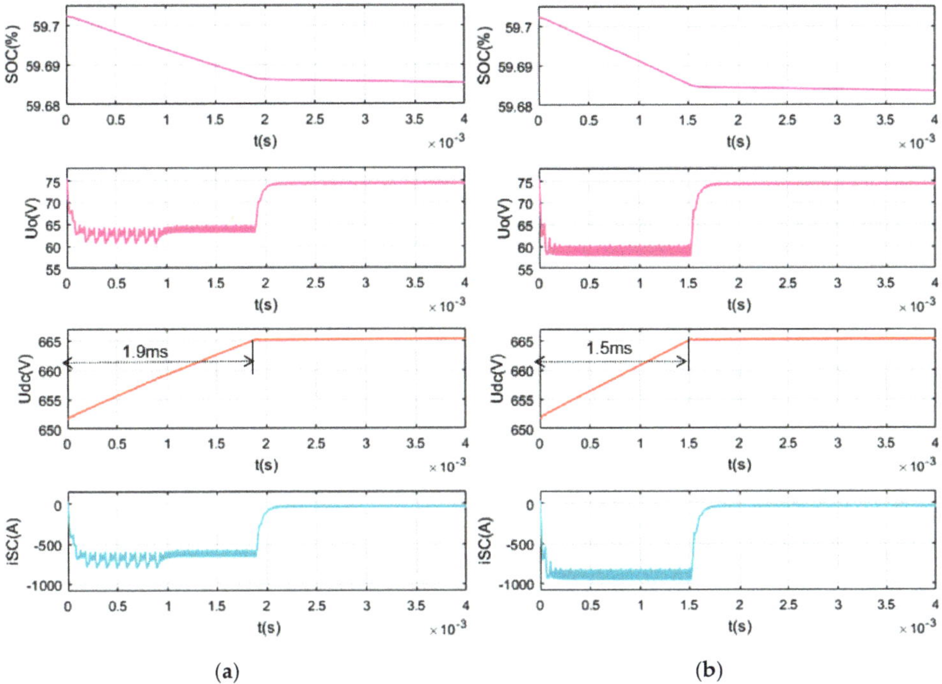

Figure 7. Simulation waveform diagram when the initial voltage, U_{dc}, is 651.7 V and the initial voltage, U_o, is 75 V: (**a**) MPC only for the DAB; (**b**) MPC for the DAB-SC system.

When the DC bus voltage, U_{dc}, is lower than the lower limit voltage of 665 V, which is 651.7 V, and the initial value of the voltage, U_o, on the SC side is 58.6 V, the dynamic response time of the system is 24.5 ms when the bus voltage is stabilized to 665 V by using the DAB converter only to establish a prediction model to control the SC discharge. The simulation waveform is shown in Figure 8a. Under the same conditions, using the overall prediction model composed of the DAB converter and the SC to control the SC discharge to stabilize the bus voltage to 665 V, the dynamic response time of the system is 7.5 ms, and the simulation waveform is shown in Figure 8b.

It can be seen from the simulation results in Figures 7 and 8 that when the SC is discharged from different SOCs, the method of using the overall model of the DAB converter and the SC for predictive control to stabilize the bus voltage always obtains faster response times than the method of using the MPC only for the DAB converter to stabilize the bus voltage.

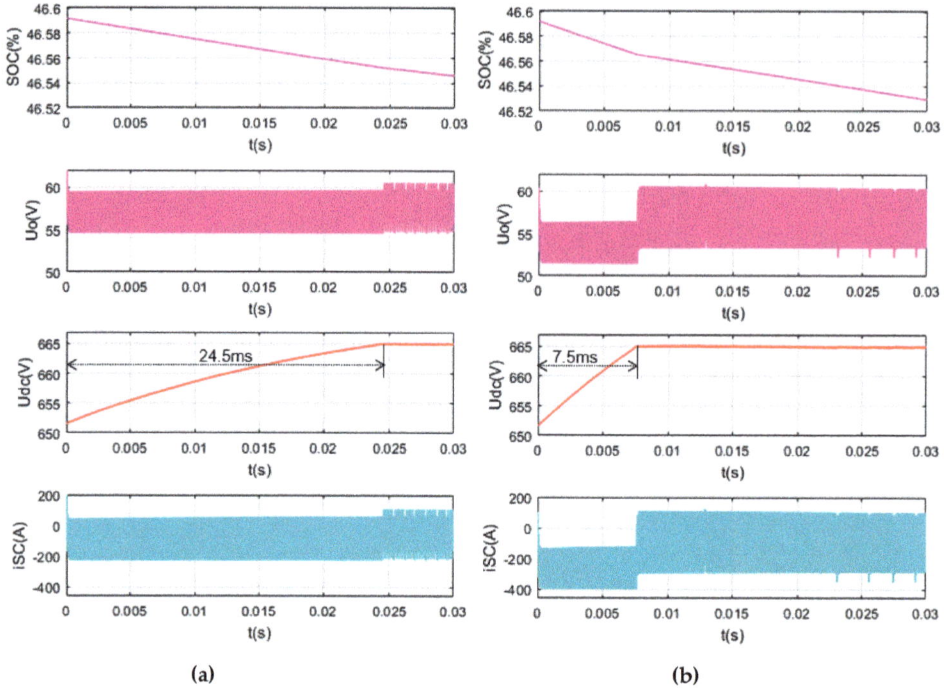

Figure 8. Simulation waveform diagram when the initial voltage, U_{dc}, is 651.7 V and the initial voltage, U_o, is 58.6 V: (**a**) MPC only for the DAB; (**b**) MPC for the DAB-SC system.

5.2. SC Charge in the Normal Working Range

When the DC bus voltage, U_{dc}, is higher than the higher limit voltage of 735 V, which is 749.7 V, and the initial value of the voltage, U_o, on the SC side is 75 V, the dynamic response time of the system is 1.7 ms when the bus voltage is stabilized to 735 V by using the DAB converter only to establish a prediction model to control the SC charge. The simulation waveform is shown in Figure 9a. Under the same conditions, using the overall prediction model composed of the DAB converter and the SC to control the SC charge to stabilize the bus voltage to 735 V, the dynamic response time of the system is 1.1 ms, and the simulation waveform is shown in Figure 9b. Under the simulation conditions, a detailed waveform of the drive signal is shown in Figure A2 (see Appendix B).

When the output power of the photovoltaic is high, the voltage, U_{dc}, on the DC bus side is higher than the higher limit voltage of 735 V, which is 749.7 V, and the initial value of the voltage, U_o, on the SC side is 93.6 V, the dynamic response time of the system is 14 ms when the bus voltage is stabilized to 735 V by using the DAB converter only to establish a prediction model to control the SC charge. The simulation waveform is shown in Figure 10a. Under the same conditions, using the overall prediction model composed of the DAB converter and the SC to control the SC charge to stabilize the bus voltage to 735 V, the dynamic response time of the system is 5.9 ms, and the simulation waveform is shown in Figure 10b.

It can be seen from the simulation results in Figures 9 and 10 that when the SC is charged from different SOCs, the method of using the overall model of the DAB converter and the SC for predictive control to stabilize the bus voltage always obtains faster response times than the method of using the MPC only for the DAB converter to stabilize the bus voltage.

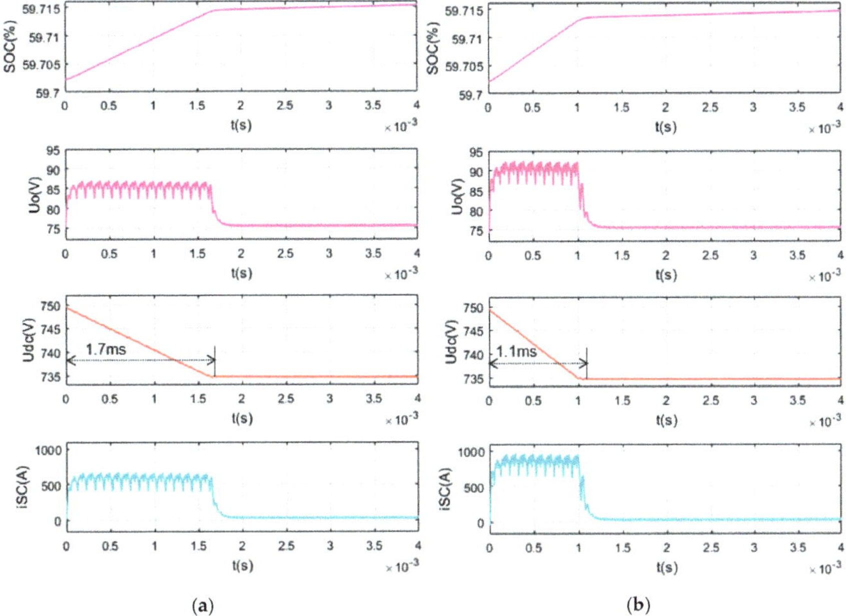

Figure 9. Simulation waveform diagram when the initial voltage, U_{dc}, is 749.7 V and the initial voltage, U_o, is 75 V: (**a**) MPC only for the DAB; (**b**) MPC for the DAB-SC system.

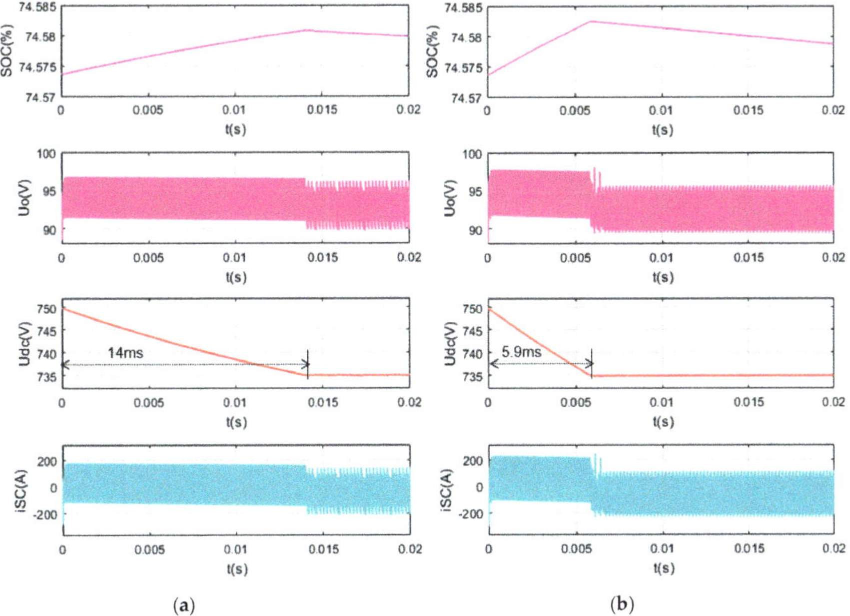

Figure 10. Simulation waveform diagram when the initial voltage, U_{dc}, is 749.7 V and the initial voltage, U_o, is 93.6 V: (**a**) MPC only for the DAB; (**b**) MPC for the DAB-SC system.

5.3. SC Discharge Closer to the Lower Limit

Through the observations of Figures 7–10, it can be found that in these four cases, compared with the method of using the MPC only for the DAB converter to stabilize the bus voltage, the method of using the overall prediction model composed of the DAB converter and the SC to stabilize the bus voltage makes the SC voltage have a larger amount of change. Figure 11 contains the simulation waveforms obtained by using two different control methods when the voltage, U_{dc}, on the DC bus side is 651.7 V and the initial value of the voltage, U_o, on the SC side is 58 V. As can be seen from Figure 11a, in this case, a method of establishing a prediction model only for the DAB converter to control the discharge of the SC cannot stabilize the bus voltage to 665 V. It can be seen from Figure 11b that the dynamic response time of the system is 12 ms by using the overall prediction model composed of the DAB converter and the SC to control the SC discharge to stabilize the bus voltage to 665 V.

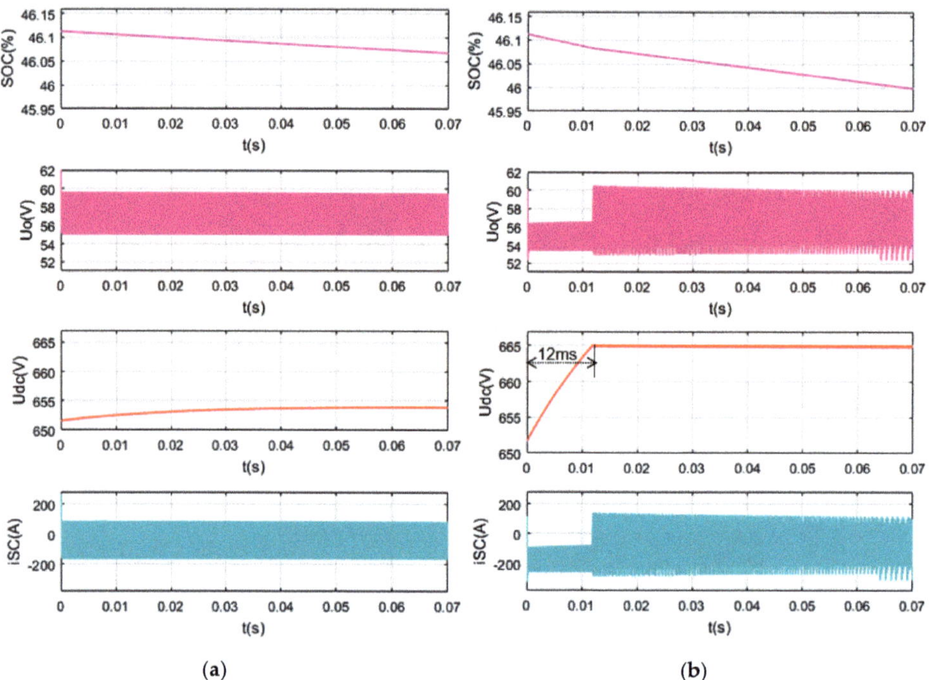

Figure 11. Simulation waveform diagram when the initial voltage, U_{dc}, is 651.7 V and the initial voltage, U_o, is 58 V: (a) MPC only for the DAB; (b) MPC for the DAB-SC system.

When the voltage, U_{dc}, on the DC bus side is 651.7 V, and the initial value of the voltage, U_o, on the SC side is 56.58 V, using the overall prediction model composed of the DAB converter and the SC to control the SC charge to stabilize the bus voltage to 665 V, the dynamic response time of the system is 15 ms, and the simulation waveform is shown in Figure 12.

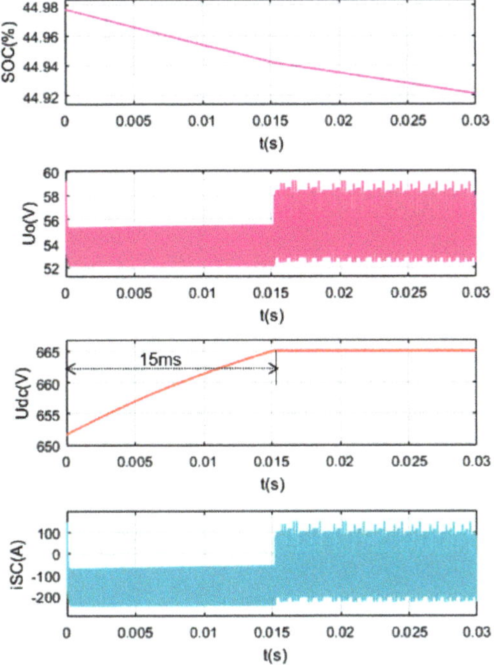

Figure 12. Simulation waveform diagram of the MPC for the DAB-SC system when the initial voltage, U_{dc}, is 651.7 V and the initial voltage, U_o, is 56.58 V.

5.4. SC Charge Closer to the Upper Limit

Figure 13 contains the simulation waveforms obtained by using two different control methods when the voltage, U_{dc}, on the DC bus side is 749.7 V and the initial value of the voltage, U_o, on the SC side is 94 V. As can be seen from Figure 13a, in this case, the method of establishing a prediction model only for the DAB converter to control the charge of the SC cannot stabilize the bus voltage to 735 V. It can be seen from Figure 13b that the dynamic response time of the system is 6 ms by using the overall prediction model composed of the DAB converter and the SC to control the SC charge to stabilize the bus voltage to 735 V.

When the voltage, U_{dc}, on the DC bus side is 749.7 V, and the initial value of the voltage, U_o, on the SC side is 98.1 V, using the overall prediction model composed of the DAB converter and the SC to control the SC charge to stabilize the bus voltage to 735 V, the dynamic response time of the system is 22.3 ms, and the simulation waveform is shown in Figure 14.

It can be seen from the simulation results of Figures 11–14 that compared with the method of establishing a prediction model only for the DAB converter to stabilize the bus voltage, the method of using the overall prediction model composed of the DAB converter and the SC to stabilize the bus voltage increases the actual upper limit of the SC voltage, reduces the actual lower limit of the SC voltage, and expands the actual capacity utilization range of the SC by 18.63%.

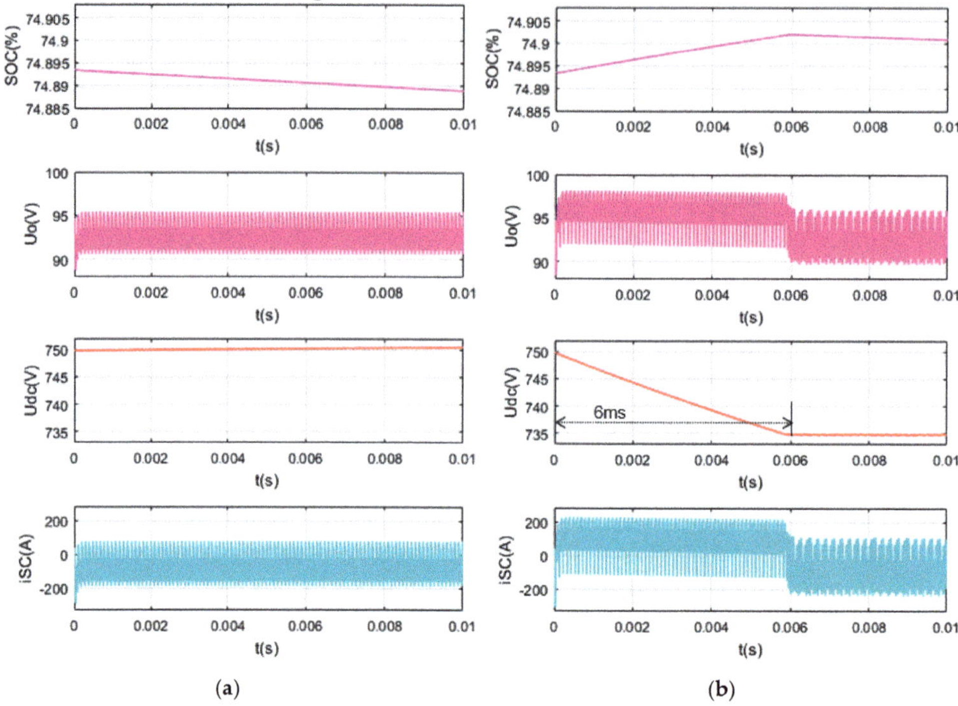

(a)   (b)

Figure 13. Experimental waveform diagram when the initial voltage, U_{dc}, is 749.7 V and the initial voltage, U_o, is 94 V: (**a**) MPC only for the DAB; (**b**) MPC for the DAB-SC system.

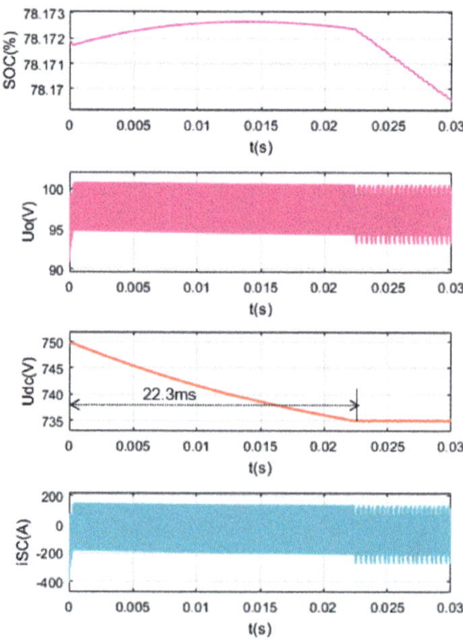

Figure 14. Simulation waveform diagram of the MPC for the DAB-SC system when the initial voltage, U_{dc}, is 749.7 V and the initial voltage, u_o, is 98.1 V.

6. Conclusions

In this paper, a hybrid MPC strategy composed of a DAB and SC was proposed in order to improve the dynamic characteristics of an energy storage system. The structural characteristics of the SCs were analyzed, and a suitable equivalent model of the SC was chosen for its fast response characteristics. The working principle of the DAB converter in each stage under single-phase shift control was analyzed and the prediction model of the DAB-SC system was established. By combining the SC model and the DAB model, an appropriate evaluation target was selected to establish the objective function for model prediction. A simulation model of the DAB-SC system was built in MATLAB/Simulink for verification. The simulation results show that the proposed method with the overall prediction model composed of a DAB converter and SC can effectively improve the overall response speed of the system and expand the actual capacity utilization range of the SC. The validity of the proposed method was verified and the proposed method has good application prospects in improving the dynamic response performance of energy storage systems.

Author Contributions: Conceptualization, J.G. and L.W.; methodology, J.G.; software, J.G.; validation, J.G., L.W. and T.W.; formal analysis, C.X.; investigation, H.L.; resources, C.X.; data curation, J.G.; writing—original draft preparation, J.G.; writing—review and editing, J.G. and L.W.; visualization, J.G.; supervision, L.W.; project administration, T.W.; funding acquisition, L.W.

Funding: This research was funded by National Natural Science Foundation Youth Project of China, grant number 51607060.

Conflicts of Interest: The authors declare no conflict of interest.

Abbreviation

SC	Supercapacitor
DAB	Dual active bridge
RC	Resistor-capacitor
ANN	Artificial neural network
DC	Direct current
ZVS	Zero voltage switch
AC	Alternative current
MPC	Model predictive control
SOC	State of charge

Appendix A

It can be seen from Figure A1 that the secondary side drive signal of the converter leads the primary side drive signal in SC discharge state. And comparing Figures A1a and A1b, it can be seen that compared with the method of establishing a prediction model only for the DAB converter to control SC discharge, the method of controlling SC discharge by using the overall prediction model composed of the DAB converter and SC has a larger amplitude of inductor current, and due to the internal resistance R_{ES} of SC, the voltage on the low voltage side declines instantaneously when SC is discharged, making $U_{dc} > nU_o$, therefore, the inductor current in the power transfer mode of the converter decreases linearly when SC is discharged.

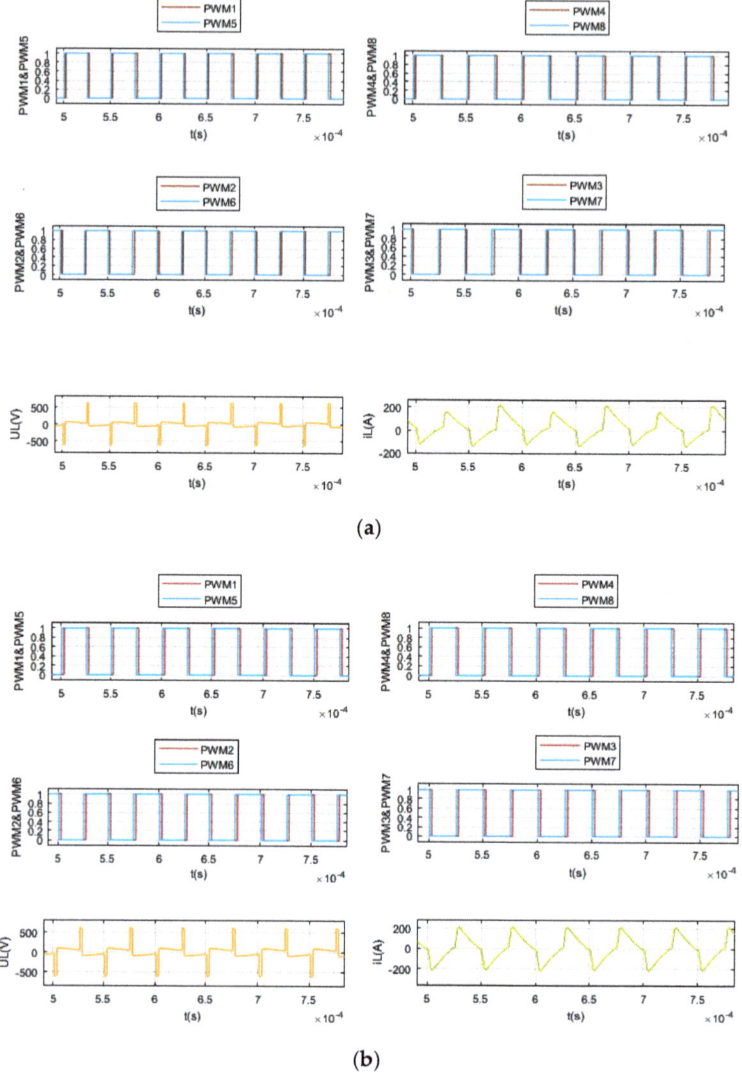

Figure A1. Detailed simulation waveform diagram of the drive signal when U_{dc} initial voltage is 651.7 V and U_o initial voltage is 75 V: (**a**) MPC only for DAB; (**b**) MPC for DAB-SC system.

Appendix B

It can be seen from Figure A2 that the primary side drive signal of the converter leads the secondary side drive signal in SC discharge state. And comparing Figures A2a and A2b, it can be seen that compared with the method of establishing a prediction model only for the DAB converter to control SC discharge, the method of controlling SC discharge by using the overall prediction model composed of the DAB converter and SC has a larger amplitude of the inductor current, and due to the internal resistance R_{ES} of SC, the voltage on the low voltage side rises instantaneously when SC is charged, making $U_{dc} < nU_o$, therefore, the inductor current in the power transfer mode of the converter decreases linearly when SC is charged.

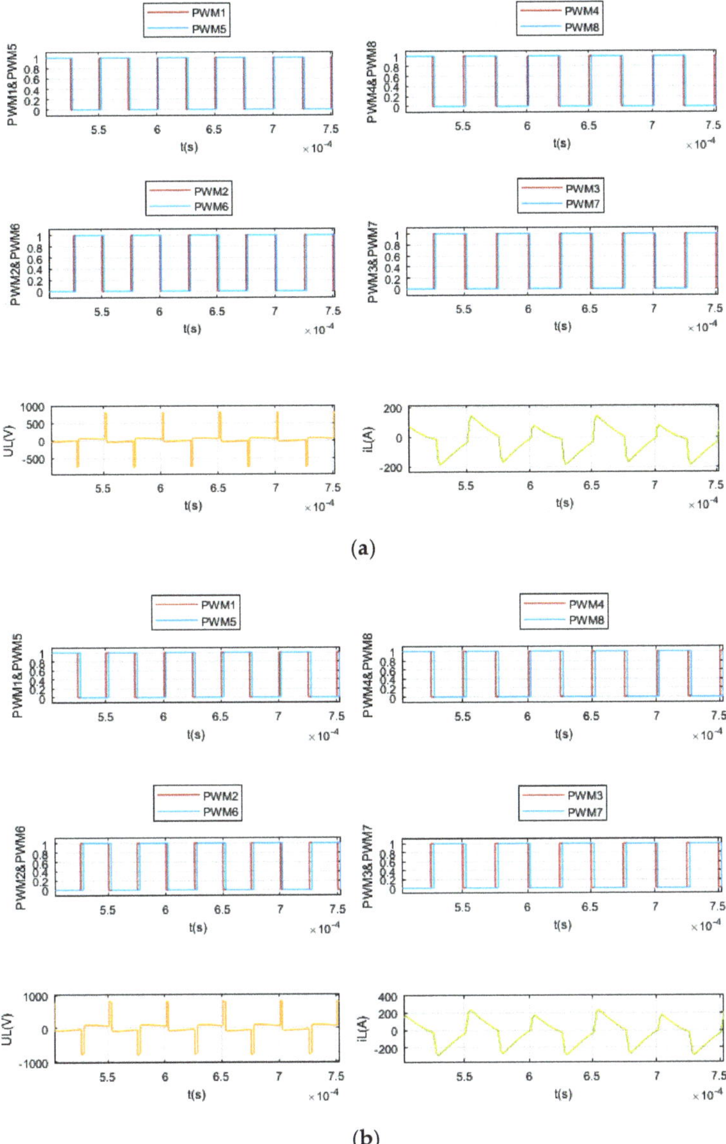

Figure A2. Detailed Simulation waveform diagram of the drive signal when U_{dc} initial voltage is 749.7 V and U_{o} initial voltage is 75 V: (**a**) MPC only for DAB; (**b**) MPC for DAB-SC system.

References

1. Liu, J.; Zhang, L. Strategy Design of Hybrid Energy Storage System for Smoothing Wind Power Fluctuations. *Energies* **2016**, *9*, 991. [CrossRef]
2. Wang, L. Research of Converter Topology Based on Super Capacitor Energy Storage System. *Electron. Des. Eng.* **2018**, *26*, 29–133.
3. Xun, Q.; Wang, P.; Quan, L.; Xu, Y. Voltage Balancing Strategy Based on Cooperative Control for Super Capacitor. *Power Electron.* **2017**, *51*, 113–116.

4. Han, W.; Huang, S.; Dai, Q.; Wan, J. Modeling and Simulation of an Elevator Energy-saving System Based on Isolated Bidirectional DC/DC Converter. *Electr. Autom.* **2015**, *37*, 7–9.

5. Lv, X.; Liu, C.; Wang, Y.; Li, Z. Research of Bidirectional DC-DC Converter in Distributed Wind-photovoltaic-storage System. *Electr. Meas. Instrum.* **2017**, *54*, 16–20, 41.

6. Choi, W.; Rho, K.M.; Cho, B.H. Fundamental Duty Modulation of Dual-Active-Bridge Converter for Wide Range Operation. *IEEE Trans. Power Electron.* **2016**, *31*, 4048–4064. [CrossRef]

7. Sha, D.; Wang, X.; Chen, D. High Efficiency Current-Fed DAB DC–DC Converter with ZVS Achievement throughout Full Range of Load Using Optimized Switching Patterns. *IEEE Trans. Power Electron.* **2018**, *33*, 1347–1357. [CrossRef]

8. Takagi, K.; Fujita, H. Dynamic Control and Performance of a Dual-Active-Bridge DC-DC Converter. *IEEE Trans. Power Electron.* **2018**, *33*, 7858–7866. [CrossRef]

9. Yi, L.; Chen, Y.; Zhang, H.; Guo, Z. Unified Control for Mode Swap of Bidirectional DC/DC Converter in Distributed Generation System. *Power Electron.* **2017**, *51*, 4–7.

10. Oggier, G.; Ordonez, M.; Galvez, J.; Luchino, F. Fast Transient Boundary Control and Steady-state Operation of the DAB Converter Using the Natural Switching Surface. *IEEE Trans. Power Electron.* **2014**, *29*, 946–957. [CrossRef]

11. Hou, N.; Song, W.; Wu, M. A Load Current Feedforward Control Scheme of DAB DC/DC Converters. *Proc. Csee* **2016**, *36*, 2478–2485.

12. Song, W.; Hou, N.; Wu, M. Virtual Direct Power Control Scheme of DAB DC-DC Converters for Fast Dynamic Response. *IEEE Trans. Power Electron.* **2018**, *33*, 1750–1759. [CrossRef]

13. Yang, K.; Song, W.; An, F.; Hou, N. Rapid Dynamic Response Control Method of Dual-active-bridge DC-DC Converters in Current Source Mode. *Proc. Csee* **2018**, *38*, 2439–2447, S23.

14. An, F.; Song, W.; Yang, K. Model Predictive Control and Power Balance Scheme of Dual-active-bridge DC-DC Converters in Power Electronic Transformer. *Proc. Csee* **2018**, *38*, 3921–3929, S22.

15. Wang, X.; Han, M.; Gao, F.; Wang, M.; Wang, K. A Review of Modeling in New Energy Research on Super-capacitor. *World Power Supply* **2016**, *1*, 37–40, S28.

16. Yu, P.; Jing, T.; Yang, R. Capacitance Measurement and Parameter Estimation Method for SCs with Variable Capacitance. *Trans. Chin. Soc. Agric. Eng.* **2016**, *32*, 169–174.

17. Hu, B.; Yang, Z.; Lin, F.; Zhao, K.; Zhao, Y. Equivalent Circuit Model of Super Capacitor Group for Urban Rail Transit Application. *Electr. Drive Locomot.* **2013**, *5*, 65–68, 74.

18. Feng, J.; Xu, F.; Li, L.; Qin, Y.; Shi, X. Capacitance Measurement of SCs Based on Discharge Procedure with Constant Load. *Electron. Compon. Mater.* **2016**, *35*, 88–91.

19. Yuan, X. A Research on Phase-shift-control of the Full-bridge-isolated DC/DC Converter. *Electr. Autom.* **2017**, *39*, 7–10.

20. Kan, J.; Wu, Y.; Xie, S.; Tang, Y.; Xue, Y.; Yao, Z.; Zhang, B. Dual-active-bridge Bidirectional DC/DC Converter Based on High-frequency AC Buck-Boost Theory. *Proc. Csee* **2017**, *37*, 1797–1807, S24.

21. Xue, Y.; Zhou, J.; Cui, Y. Simulation Research on Coordinated Control Strategy of Hybrid Energy Storage Based on DC Microgrid. *Electr. Mach. Control Appl.* **2017**, *44*, 19–25, 37.

Article

Design and Validation of Ultra-Fast Charging Infrastructures Based on Supercapacitors for Urban Public Transportation Applications

Fernando Ortenzi [1],* , Manlio Pasquali [1], Pier Paolo Prosini [1], Alessandro Lidozzi [2] and Marco Di Benedetto [2]

[1]   ENEA - Italian Agency for New Technologies, Energy and Sustainable Economic Development, 00123 Rome, Italy; manlio.pasquali@enea.it (M.P.); pierpaolo.prosini@enea.it (P.P.P.)
[2]   ROMA TRE University, Department of Engineering, 00146 Rome, Italy; alessandro.lidozzi@uniroma3.it (A.L.); marco.dibenedetto@uniroma3.it (M.D.B.)
*   Correspondence: fernando.ortenzi@enea.it; Tel.: +39-6-30486184

Received: 20 May 2019; Accepted: 17 June 2019; Published: 19 June 2019

Abstract: The last few decades have seen a significant increase in the number of electric vehicles (EVs) for private and public transportation around the world. This has resulted in high power demands on the electrical grid, especially when fast and ultra-fast or flash (at the bus-stop) charging are required. Consequently, a ground storage should be used in order to mitigate the peak power request period. This paper deals with an innovative and simple fast charging infrastructure based on supercapacitors, used to charge the energy storage system on board electric buses. According to the charging level of the electric bus, the proposed fast charging system is able to provide the maximum power of 180 kW without exceeding 30 s and without using DC–DC converters. In order to limit the maximum charging current, the electric bus is charged in three steps through three different connectors placed between the supercapacitors on board the bus and the fast charging system. The fast charging system has been carefully designed, taking into account several system parameters, such as charging time, maximum current, and voltage. Experimental tests have been performed on a fast charging station prototype to validate the theoretical analysis and functionality of the proposed architecture.

Keywords: supercapacitors; fast and flash charging; urban public transportation

---

1. Introduction

The electric vehicles (EVs) in public transportation could be the keystone to limit a city's traffic problem, noise level, and pollution within urban complexes with high population density [1,2]. In fact, having electric public vehicles in the city center will have a large positive environmental effect on both noise and pollution. According to the many advantages provided by EVs, more and more cities have been replacing their public vehicles with electric vehicles [3–8]. In [7], a cost-benefit method was employed for the scheduling of an electric city bus fleet on a single route. Three different charging infrastructure scenarios were considered. In the first scenario, only one charging station was used. The second scenario considered two charging stations that were located at the same terminus. In the third scenario, two charging stations were located at opposite terminuses. The first scenario is the cheaper solution compared to the others. The second solution is more suitable in terms of energy demand and traffic conditions. However, the electric vehicles, such as electric buses, are limited by the onboard storage system and the long charging times, which depends on the charging infrastructure. There are several different methods to charge the electric buses [8,9].

1.   Overnight charging method. In this case, the electric buses are equipped with a huge energy storage system in order to satisfy the bus routes during the day. The large storage system (i.e.,

for a 12 m bus there are about 350 kWh of lithium batteries) must be recharged during the night (in about 6 h).

2. Route charging method. In this method, the buses are charging in the bus-station during each break at the end of the routes. The charging time takes only 10–15 min. Compared to the previous method, the energy storage system is significantly reduced (i.e., for a 12 m bus there are about 40–60 kWh of lithium batteries), whereas the charging infrastructure is more stressed than the overnight charger due to the higher power requests. Thus, in this case, the advantage is the smaller energy storage system onboard the buses, leading to less costs.

3. Flash charging method. In this case, the electric bus recharges at bus stops in just a few seconds. In fact, the charging time takes only 20–30 s (when the passengers are getting on the bus). Then, each time the buses come to the final stop throughout the day, it quickly recharges. The energy storage system on board the bus is smaller than in previous cases (i.e., for a 12 m bus there are about 1–2 kWh of lithium batteries). Since the charge/discharge are carried out at high power, the energy storage system is composed of supercapacitors instead of batteries. Consequently, the supercapacitors are used to realize charging infrastructure. Using this method, the costs are drastically reduced due to the smaller energy storage system on board the bus compared to the previous methods.

Compared to the slow charging stations, the main advantages of the fast charging stations are an extended range autonomy and a smaller storage system. Unfortunately, the connection of the unexpected loads, such as an energy storage system, to the electrical grid may cause unwanted effects on the grid due to the great power demands for short times. For instance, the Hess-ABB Tosa (Trolleybus Optimisation System Alimentation), an 18 m bus fitted out by ABB with an optimized complete traction chain, is composed by 40 kWh energy-storage units and an automatic energy transfer system at bus-stops. This electric bus is charged two times on the 18 km route. The first charging point, called Palexpo Flash Station, is equipped with a 400 kW power supply unit and the charging time is only 15 s, whereas the second charging point is composed by a 200 kW power supply unit, and the charging time is 3–4 min. These storage systems inevitably require expensive components with a large effect on investment costs. Performance and cost optimization of charging stations and electric storage systems pass through technology solutions able to integrate devices with different functions.

In [1], the evaluating methodology of power-quality field measurements of an electric bus charging station in Taiwan was proposed. In this case, the charging station under study had 11 charging poles, including two fast charging and nine slow charging poles. Experimental measurements of voltage flickers, harmonic distortion, including electrical power measurements, and three phase voltages were performed. In [2], a control structure for the DC charging station was proposed. The charging station makes use of the frontend AC–DC converter, including the power grid filters, and backend DC–DC converter connected via DC link capacitor. The DC–DC converter is controlled through the P-controller, whereas the AC–DC converter is controlled using the PI (Proportional-Integral) controller, where the input error is provided by the difference between the measured grid voltage and reference voltages. The charging station proposed in [5] is equipped with a modular multilevel converter (MMC) and the control structure is based on finite control set-model predictive control.

Such charging processes can be realized using the supercapacitors. The advantages of supercapacitors can be highlighted in three different characteristics [10]: (1) high specific power—supercapacitors can be totally discharged and charged in the order of seconds; (2) long life, to the order of 106 cycles; (3) low specific energy.

The use of supercapacitors in both hybrid vehicles and hybrid storage configurations, including microcars or forklifts, has been discussed in different papers [11–13]. In an electric microcar tested on a dynamometer chassis running the urban part of the previous type approval procedure driving cycle (ECE15) [11], the addition of supercapacitors to the battery storage brought a reduction of the maximum current from lead-acid batteries and then advantages in terms of vehicle range (more than 50%) and battery life (more than double). The same results have been obtained also on a forklift [12].

Such advantages (in terms of range) have not been obtained if the vehicle is equipped with lithium-ion batteries due to their lower internal resistance [13].

In this paper, a flash charging infrastructure, a three-year project called "Ricerca di Sistema Elettrico", has been designed, evaluated in terms of investment costs [14,15], and tested using a reduced scale model prototype [16]. Such charging infrastructure needs to be equipped with a storage system based on supercapacitors due to the high-power transfer. The realized electric bus has a hybrid storage system composed of lead acid batteries for the auxiliaries and backup, as well as the supercapacitors for the traction. Furthermore, the proposed system presents an innovative concept, where the charging is directly performed between two supercapacitors without use of the converter: the first one placed in the charging station and the second one placed on board the electric bus. To limit the charging current, an inductor is connected between the supercapacitors and the charge is made in three different configurations by changing the connection of the supercapacitors in the charging station. The advantages of the present configuration are that the charging station is simpler and does not have any expensive devices like a converter; the only components used are contactors and an inductance, cheaper than a DC–DC converter designed for such a purpose.

The paper is organized in six parts. The background of the charging station in the electric buses is discussed in Section 1. The electric bus details (Tecnobus Gulliver U500), equipped with the pantograph and the hybrid storage system, adapted to be compliant with the present charging system, is explained in Section 2. The charging architecture and the charging control strategy is discussed in Section 3, while experimental results are presented in Section 4. Section 5 investigates the possible structure and the management of a complex charging network for public transport. Finally, Section 6 gives the conclusions.

2. The Pantograph Equipped Vehicle

The vehicle used to test the flash charge infrastructure was a minibus, Tecnobus Gulliver U500, also used for previous fast charge experimentation [17,18]. The specifications of the minibus are reported in Table 1.

Table 1. Tecnobus Gulliver U500 Specification.

Parameter	Stock Version	Flash Charge Version
Length		5.3 m
Passengers		28
Max Speed		33 km/h
Electric Motor		235 Nm@950 Rpm 24.8 kW@1035 Rpm
Auxiliaries consumption		~0.9 kW
Weight (kg)	4285 (Tare)–6045 (Gross)	3500 (Tare)–6045 (Gross)
Storage	Lead Acid 72V/585 Ah (42.1 kWh)	Lead Acid: 72V/120 Ah (8.6 kWh) Supercapacitors: 375V 63F (410 Wh)

This vehicle was originally equipped with lead acid batteries. The previous experimental results, carried out at the ENEA (Italian Agency for New Technologies, Energy and Sustainable Economic Development) Casaccia Research Center, were also obtained using the LFP (lithium iron phosphate) batteries [18].

The energy storage system on board the bus was specifically designed in order to use the flash charging method. The energy storage was composed by a hybrid system, where 8.6 kWh lead acid batteries (72 V/120 Ah) were used together with three Maxwell supercapacitor modules (125 V, 63 F) connected in series, as shown in Figure 1a. The lead acid batteries were used for both the auxiliaries and as a backup battery, when the supercapacitors were totally discharged during the route of the bus. Furthermore, the batteries helped to partially discharge the supercapacitors, when their voltage was too high to be charged. The nominal energy of the supercapacitors was 410 Wh; nevertheless, since the supercapacitors worked at half their nominal voltage, only 75% of the supercapacitors' nominal energy was used. The description of the architecture on board the bus is explained in more detail in [19]. Figure 1b shows the pantograph installed on the bus. The pantograph was a fast charging device manufactured by Schunk with the following characteristics:

- Maximum operating voltage: 1000 V;
- Continuous charging current: 150 A (6 h);
- Maximum charging current: 1000 A (15 min);
- Service life: 100,000 cycles.

(a)   (b)

Figure 1. (**a**) Hybrid Storage system on board the bus; (**b**) and the pantograph for ultrafast charge.

3. Charging Architecture

The main characteristics of the flash charging method proposed in this paper are the following:

(1) The charging time at every bus-stop must be less than the pick-up and drop-off times of the passengers. The acceptable charging time is fixed at 30 s;

(2) The charging station supercapacitors must be rapidly charged. A maximum charging time is set at 3 min;

(3) The charging station must charge the electric bus even if the bus storage system has a different voltage from the designed value. The bus energy consumption is not constant during its route, depending on traffic and passenger load. Consequently, the charging station must be able to charge the bus independently of the state of the charge of the onboard supercapacitors;

(4) A ground storage is required to reduce the peak power demanded of the power grid.

In [14–16] the charging station was designed and tested in a reduced scale prototype. In that case, the power levels were not taken into account and the electric scheme of the charging station was created to optimize the simulation model. Such a system was composed by six ultra-capacitors in the ground charging station and three onboard the vehicle, each one of 16 V and 250 F. Starting from this analysis, in this paper a real scale protype was built. In this paper, the proposed charging station was built with the architecture explained below.

The block scheme of the proposed charging station is depicted in Figure 2. Five contactors were used to operate the charging process. The charging was made in three steps: (1) four modules, connected two in series and two in parallel (2s2p), were used to charge the vehicle; (2) an extra module was added in series to the others; and (3) a sixth module was connected in series to the others.

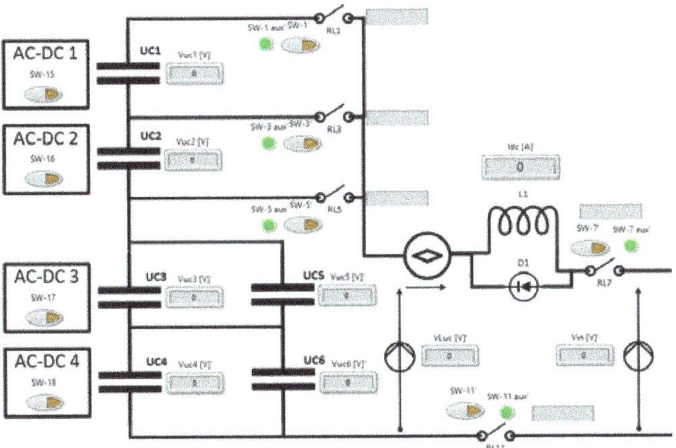

Figure 2. Block scheme of the charging station.

In the first step, the contactors RL5, RL11, and RL7 were closed to allow the four-supercapacitor module to transfer energy from the charging station to the onboard supercapacitors. When the charging transfer was almost complete (or the current fell below 20 A), the contactor RL5 was opened, while the contactor RL3 was closed to start the second step. When the second step was finished, the closure of the RL1 started the third step. Finally, four AC–DC converters with 2 kW rated power were used to charge the supercapacitors after the energy transfer process. The maximum charging current was limited, using the inductor and resistor with the parameters listed in Table 2. The inductor is shown in Figure 3. The supercapacitor configuration together with the inductance is an efficient strategy to manage the flash charge in a cheap way. The reason for the use of three recharge steps instead of one is related to the possibility of reducing the maximum recharge current and optimizing the energy transfer.

Figure 3. Inductor used in the charging station.

Table 2. Inductor and resistor parameters.

L = 5.4 mH
Resistance = 0.155 Ω
Maximum Current = 500 A
K_u = 0.5
I_{RMS} = 187 A

4. Results

Different tests were carried out in order to validate the design of the proposed charging station.

- Verifying that the charging process is completed within a prefixed time: At this stage the charging process is non-automatized, so from one step to the following one there could be a delay that could be easily removed in the future. The end of each step and the start of the following one is set to when the current falls below a threshold value of 20 A.
- Verifying that the ground supercapacitors are fully recharged in a fixed period of time in order to be ready to charge the next bus.
- Verifying that the maximum current and voltage do not exceed the nominal ones: every supercapacitor module has a nominal voltage of 125 V, but only half of this value is used. Hence, the maximum charge voltage is 375 V, while the minimum voltage is 187.5 V
- Verifying that the charging station is able to charge the bus even if the bus storage has a supercapacitor voltage different from the minimum allowed. The bus energy consumption is not constant during the route, depending on traffic, path, and pay-load. Accordingly, the charging station must be able to charge the onboard energy storage even if it is not completely discharged.

The simulation results performed in one step (1-step) and three steps (3-step) recharge processes for the proposed system are shown in Figures 4 and 5. The 1-step charging is faster, but the 3-step charging is more efficient and the correspondent current is lower. Table 3 shows the comparison between the two charging methods.

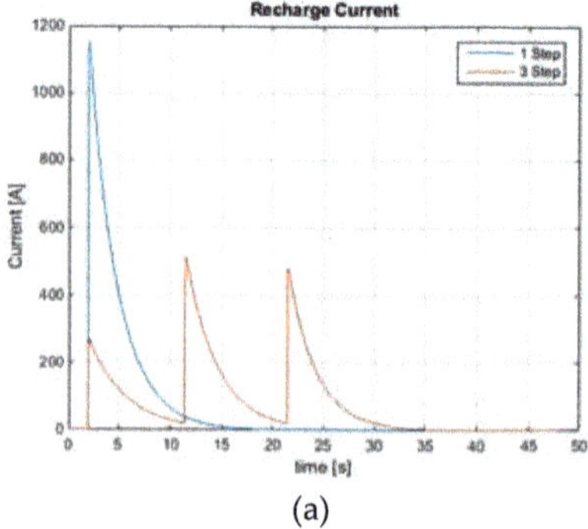

(a)

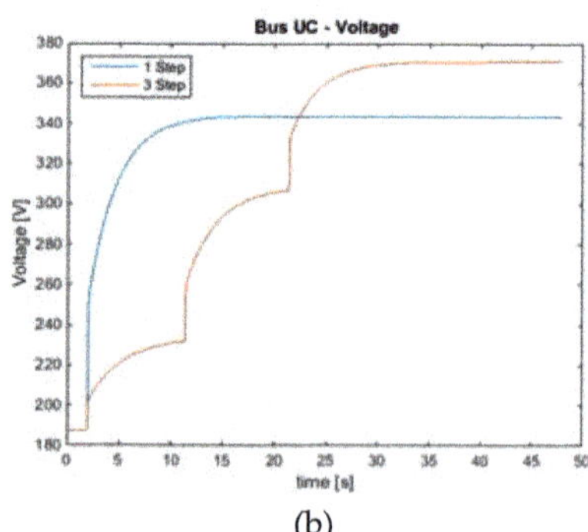

(b)

Figure 4. Comparison of 1-step and 3-step charge. (**a**) Charging current as a function of time, (**b**) bus supercapacitors voltage.

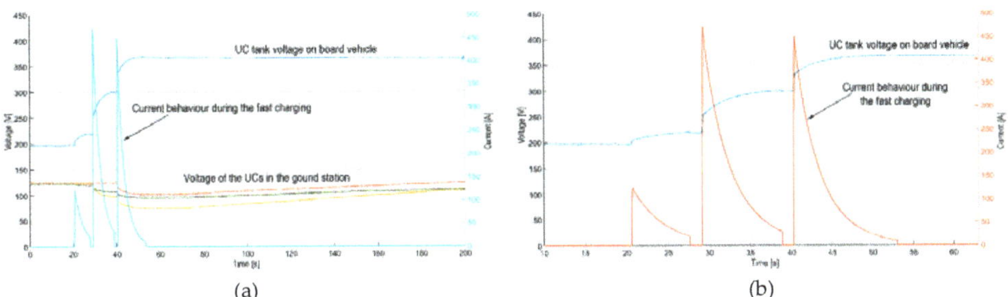

Figure 5. Voltages and currents as a function of time: (**a**) station side, (**b**) onboard side.

Table 3. Comparison of 1 and 3 step recharge.

Number Step	Recharge Time [s]	Max Current [A]	Vcbus Final [V]	ΔEuc Bus [kWh]
1	24	1152	343.7	43.58
3	40	513	370.7	53.69

It has been found that the test results depend on the initial voltage of the supercapacitors. In Table 2 the extreme conditions of recharge are considered. The voltage and current profile during a charging are shown in Figure 5. The six modules connected as in Figure 2 present different voltage values during the charge. It can be seen from Figure 5 that the three steps can be easily identified, observing three different peak current and the three different voltages (supercapacitor voltages on board the vehicle and the voltage on the ground station).

The maximum peak current, more than 400 A, was obtained during the second step and the supercapacitor module connected in the second step reached the lowest value of voltage at the end of the charge process. The charging time, without the delay times due to the manual drive of the charging steps, was within the 30 s. The charging time of the ground supercapacitors was about 2.5 min.

The charging process when the initial voltage of the onboard supercapacitors was changed from 190 V to 250 V is depicted in Figures 6 and 7. With reference to different onboard voltages, the transfer energy was reduced during the first step due to the different voltage between ground and onboard storage. The current changed from 300 A (for voltage = 190 V) to 220 A (voltage = 230 V). Additionally, the first current step is missing when the onboard voltage is 250 V (Figure 7b).

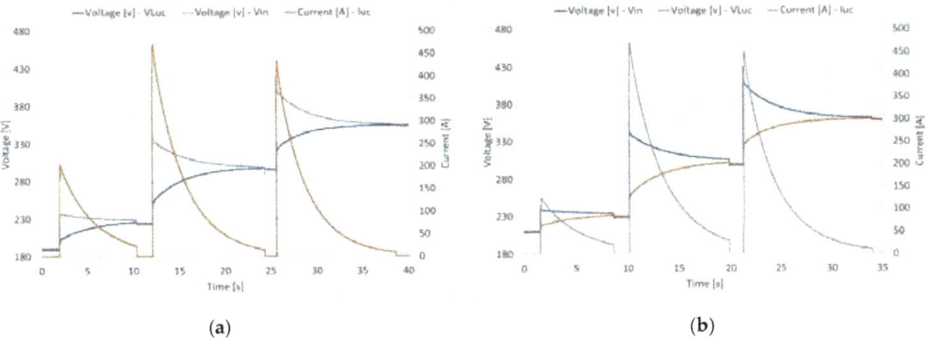

Figure 6. Charging process with on board initial voltage of: (**a**) 190 V, (**b**) 210V.

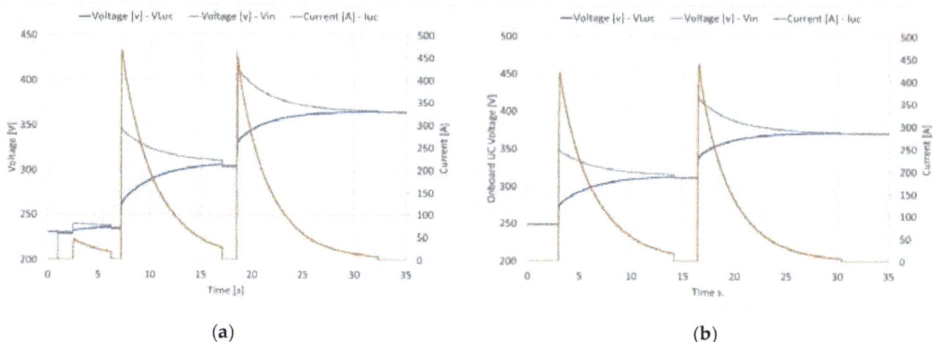

(a)   (b)

Figure 7. Charge process with onboard initial voltage of: (**a**) 230 V, (**b**) 250V.

Thanks to the lighter storage system compared to the original version (powered with 43 kWh lead acid batteries) the unladen weight passed from 4300 kg to about 3500 kg and the gross laden weight from 6300 to 5600 kg. Consequently, the range on a reference circuit, internal to the Enea Casaccia Research Center of about 0.7 km, may range from 320 Wh/km up to 450 Wh/km, depending on the number of passengers transported.

The bus route was performed in different conditions: a full and empty bus. Moreover, the state of charge (SOC) of the supercapacitors was changed from 200 V to 275 V. Figure 8 shows the reference speed profile and the SOC for different bus conditions and different voltages of the supercapacitor tank. As can be seen from Figure 8, the charging station is able to charge the onboard supercapacitors independently of their initial voltage.

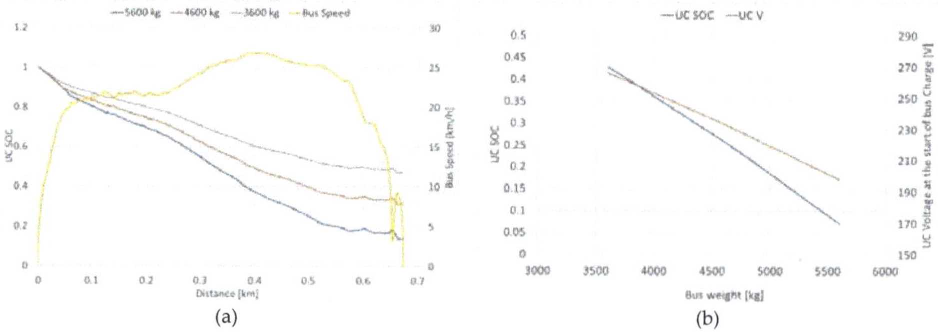

(a)   (b)

Figure 8. (**a**) Vehicle speed profile and state of charge (SOC) for different bus weights, (**b**) voltage of the supercapacitor tank and related SOC.

Figure 9 shows the final voltage at the end of the charge process as a function of the initial voltage of the supercapacitors. When the onboard SOC is close to zero, the final voltage is about 360 V, and it increases with an increase in the initial voltage. When the initial voltage reaches the maximum value (250 V), the charging station uses the two charge steps to bring the final voltage at 375 V. This is the rated voltage for the supercapacitors; it is not possible to charge the energy storage system of the bus if the initial voltage has higher values. However, the bus has a hybrid storage system composed by batteries and supercapacitors. Some of the energy from the supercapacitors is transferred to the batteries in order to decrease the supercapacitor voltage. Thus, the battery storage on board the bus can be used to discharge the supercapacitors when their voltage is too high. Furthermore, battery storage is used as a backup system when the supercapacitors are totally discharged during the route.

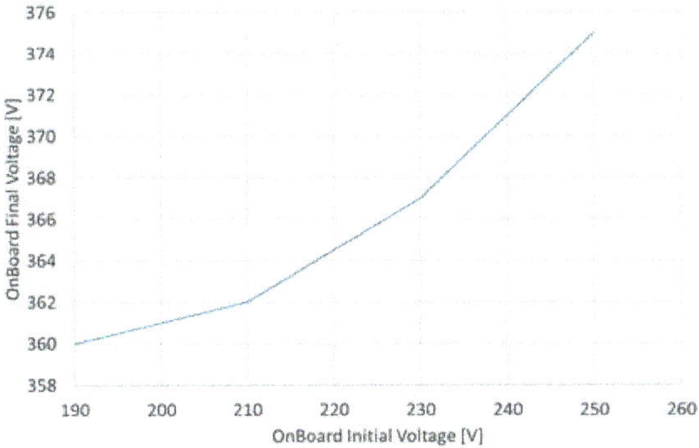

Figure 9. Final onboard supercapacitor voltage as a function of the initial voltage (as recorded before the charging process).

5. Structure of a Charging Network and its Management

In this paper, the structure of a flash charging station in the bus route was considered. Once the feasibility of the idea is verified, it is necessary to consider the charging network building for a complex transport system.

Figure 10 shows how it is possible to connect a supercapacitor bank to different bus stops, and it is also possible to connect a bus stop to different supercapacitor banks.

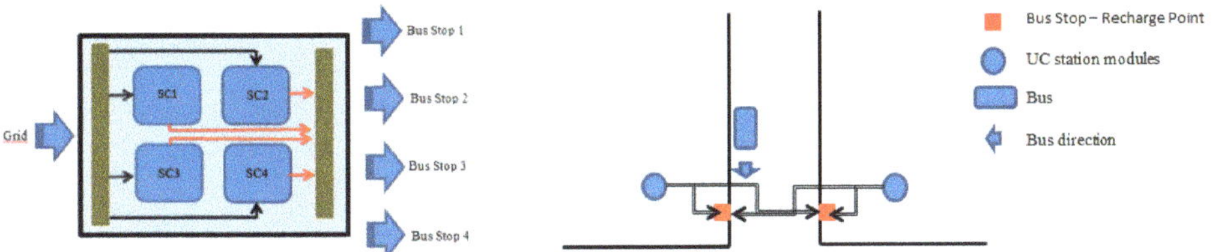

Figure 10. Proposed structure for the flash charging network.

In this way, it is possible to create a structured flash recharging network.
To build the network the following must be considered:

1. the distance between the different bus stops;
2. the frequency of passing buses;
3. the presence of transformation stations near the stops.

For example, if two stops are very close, and a large interval of time is expected for the passage of the buses, only one bank of capacitors can be used; if instead there are many stops nearby with a high frequency of passing buses it is necessary to provide more banks of capacitors, greater than the number of stops if needed.

A rule that has been established is that if two stops are more than a hundred meters apart then they are not connected.

To manage the network, it is necessary to establish the value of the supercapacitors' charge power, Figure 11 shows a possible technique for managing the charging power.

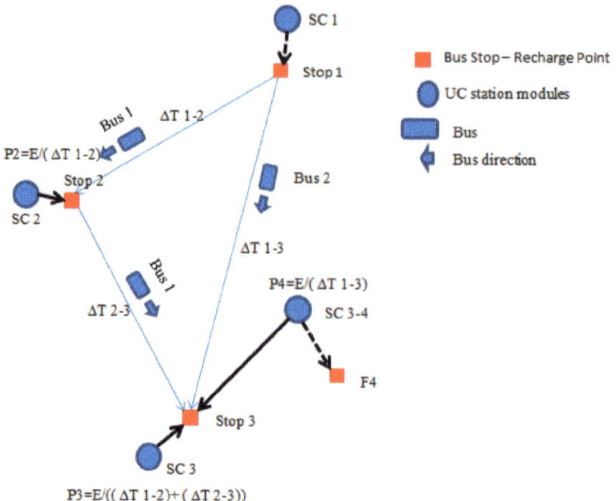

Figure 11. Proposed criteria for network management.

From the knowledge of waiting times at the bus stop, a solution for choosing the capacitor recharge power is shown in Figure 11.

The subject of study has been introduced for completeness; it will be explained in more detail in other papers, but to give an example, consider a simplified case.

Consider the following charging network (Figure 12), which can be seen as a system where the status is given by the position of the buses and the SOC of the capacitors on the ground. The input is the direction and speed of the buses, or equivalent, the waiting times at the stops and the output are the SuperCapacitorsconnections with the bus stops and their recharging powers.

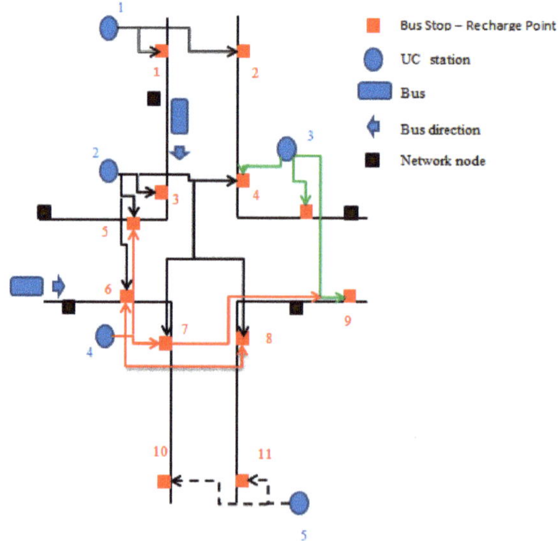

Figure 12. Example of flash charge network.

In a first example, the waiting time at the stops is considered a multiple of 3 min, the supercapacitors' recharge power is a constant and equal to 8 kW. In another case, it is possible to choose between three thresholds—8, 4, and 2.7 kW.

The following Table 4 shows the transition times at the stops for four buses.

Table 4. Bus travel times and relative stops.

Time (min)	0	3	6	9	12	15
Bus 1	N	1	N	3	N	9
Bus 2	N	6	7	10	N	N
Bus 3	N	11	N	4	2	N
Bus 4	N	N	11	N	N	9

N = No stop.

Table 5 indicates a possible management of the capacitors by adopting a single recharge threshold; the term Cif indicates that the capacitor is connected to the station f.

Table 5. Management for a single recharge threshold, Power = 8 KW.

Time (min)	C_1 (kW)	C_{1F}	C_2 (kW)	C_{2F}	C_3 (kW)	C_{3F}	C_4 (kW)	C_{4F}	C_5 (kW)	C_{5F}
0	8	1	8	6	0	0	0	0	8	11
3	0	0	0	0	0	0	8	7	8	11
6	0	0	8	3	8	4	0	0	8	10
9	8	2	0	0	0	0	0	0	0	0
12	0	0	0	0	8	9	8	9	0	0
15	0	0	0	0	0	0	0	0	0	0

Table 6 indicates a possible solution with three recharge thresholds.

Table 6. Management with three recharge thresholds, P = 2.7,4, and 8 KW.

Time (min)	C_1 (kW)	C_{1F}	C_2 (kW)	C_{2F}	C_3 (kW)	C_{3F}	C_4 (kW)	C_{4F}	C_5 (kW)	C_{5F}
0	8	1	2.7	3	2.7	4	8	6	8	11
3	2.7	2	2.7	3	2.7	4	8	7	8	11
6	2.7	2	2.7	3	2.7	4	2.7	9	8	10
9	2.7	2	0	0	4	9	2.7	9	0	0
12	0	0	0	0	4	9	2.7	9	0	0
15	0	0	0	0	0	0	0	0	0	0

Figure 13 allows the results to be compared for the same time interval.

The subject of study was introduced for completeness; it will be explained in more detail in the other papers.

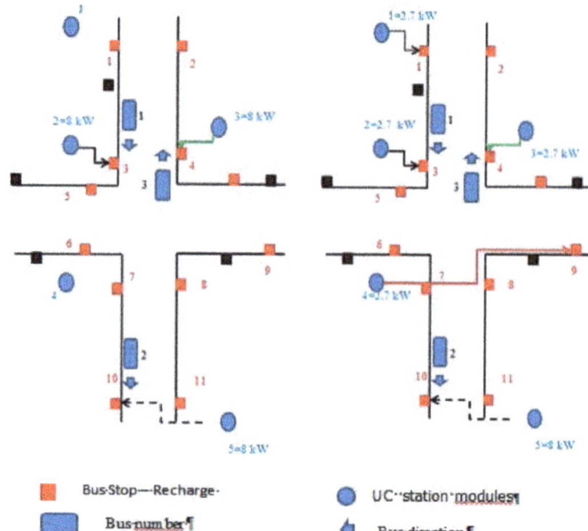

Figure 13. Bus position and configuration of the charging network for $6' < T < 9'$ in the two solutions.

6. Conclusions

The preliminary results of a flash charge system for a public transport vehicle characterized by supercapacitors and batteries storage system onboard the vehicle were presented. The charging process was divided into three steps in order to limit the maximum currents.

It can be noticed that such a charging system could be effective from both a technical and economical point of view when used for reduced distance between two adjacent stops. The necessary grid average power can be definitely fulfilled by a common urban light system due to the quite low average value. In fact, the charging process of the ground station is accomplished with a longer time between two different vehicles. The results have confirmed the theoretical, modeling, and experimental study served for sizing and implementing of the 1:1 scale system.

In the final part of the paper, suggestions have been proposed for future improvements of the charging system and for alternative structures of the flash charging stations.

Author Contributions: Funding acquisition, P.P.P.; project administration, F.O.; software, A.L., M.D.B.; writing—original draft preparation, M.P.; writing—review and editing, F.O., M.P and A.L.

Funding: This research was funded by Ministero dello Sviluppo Economico, Project C.5, Energy Storage systems for the electric system, PAR 2017.

References

1. Su, C.; Yu, J.; Chin, H.; Kuo, C. Evaluation of power-quality field measurements of an electric bus charging station using remote monitoring systems. In Proceedings of the 2016 10th International Conference on Compatibility, Power Electronics and Power Engineering (CPE-POWERENG), Bydgoszcz, Poland, 29 June–1 July 2016; pp. 58–63.
2. Varghese, A.S.; Thomas, P.; Varghese, S. An efficient voltage control strategy for fast charging of plug-in electric vehicle. In Proceedings of the 2017 Innovations in Power and Advanced Computing Technologies (i-PACT), Vellore, India, 21–22 April 2017; pp. 1–4.
3. Yu, H.; Cheli, F.; Castelli-Dezza, F. Optimal Design and Control of 4-IWD Electric Vehicles Based on a 14-DOF Vehicle Model. *IEEE Trans. Veh. Technol.* **2018**, *67*, 10457–10469. [CrossRef]
4. You, P.; Yang, Z.; Zhang, Y.; Low, S.H.; Sun, Y. Optimal Charging Schedule for a Battery Switching Station Serving Electric Buses. *IEEE Trans. Power Syst.* **2016**, *31*, 3473–3483. [CrossRef]

5. Nademi, H.; Zadeh, M.; Undeland, T. Interfacing an Electric Vehicle to the Grid with Modular Conversion Unit: A Case Study of a Charging Station and its Control Framework. In Proceedings of the IECON 2018—44th Annual Conference of the IEEE Industrial Electronics Society, Washington, DC, USA, 21–23 October 2018; pp. 5171–5176.

6. Ríos, M.A.; Muñoz, L.E.; Zambrano, S.; Albarracín, A. Load profile for a bus rapid transit flash station of full-electric buses. In Proceedings of the IEEE PES Innovative Smart Grid Technologies, Europe, Istanbul, Turkey, 12–15 October 2014; pp. 1–6.

7. Vepsäläinen, J.; Baldi, F.; Lajunen, A.; Kivekäs, K.; Tammi, K. Cost-Benefit Analysis of Electric Bus Fleet with Various Operation Intervals. In Proceedings of the 2018 21st International Conference on Intelligent Transportation Systems (ITSC), Maui, HI, USA, 4–7 November 2018; pp. 1522–1527.

8. Rahbari-Asr, N.; Chow, M. Cooperative Distributed Demand Management for Community Charging of PHEV/PEVs Based on KKT Conditions and Consensus Networks. *IEEE Trans. Ind. Inform.* **2014**, *10*, 1907–1916. [CrossRef]

9. Zhai, N.S.; Yao, Y.Y.; Zhang, D.I.; Xu, D.G. Design and Optimization for a Supercapacitor Application System. In Proceedings of the 2006 International Conference on Power System Technology, Chongqing, China, 22–26 October 2006; pp. 1–4.

10. Conti, V.; Orchi, S.; Valentini, M.P. Upgrading del sistema di supporto alle decisioni relativo all'elettrificazione di singole linee di trasporto pubblico locale. Rome, Italy, RdS/PAR2016/226. Available online: http://www.enea.it/it/Ricerca_sviluppo/documenti/ricerca-di-sistema-elettrico/adp-mise-enea-2015-2017/accumulo-di-energia/rds_par2015-193.pdf (accessed on 20 May 2019).

11. Rossi, E.; Villante, C. A hybrid city car. *IEEE Veh. Technol. Mag.* **2011**, *6*, 24–37. [CrossRef]

12. Conte, M.; Genovese, A.; Ortenzi, F.; Vellucci, F. Hybrid battery-supercapacitor storage for an electric forklift: A life-cycle cost assessment. *J. Appl. Electrochem.* **2014**, *44*, 523–532. [CrossRef]

13. Ortenzi, F.; Pede, G.; Rossi, E. Lithium-Ion/Supercapacitors Storage System Powered Microcar: Development and Testing. *Energy Procedia* **2015**, *81*, 422–429. [CrossRef]

14. Ortenzi, F.; Orchi, S.; Pede, G. Technical and economical evaluation of hybrid flash-charging stations for electric public transport. In Proceedings of the 2017 IEEE International Conference on Industrial Technology (ICIT), Toronto, ON, Canada, 22–25 March 2017; pp. 549–554.

15. Ortenzi, F.; Orchi, S.; Pede, G. Technical and Economical Evaluation of Hybrid Fast-Charging Stations for Electric Public Transport. In Proceedings of the Evs30—Electric Vehicle Symposium & Exhibition, Stuttgart, Germany, 9–11 October 2017.

16. Ortenzi, F.; Pede, M.P.G.; Lidozzi, A.; di Benedetto, M. Ultra-fast charging infrastructure for vehicle on-board ultracapacitors in urban public transportation applications. In Proceedings of the EVS31—The 31st International Electric Vehicles Symposium & Exhibition & International Electric Vehicle Technology Conference 2018, Kobe, Japan, 30 September–3 October 2018.

17. di Rienzo, R.; Baronti, F.; Vellucci, F.; Cignini, F.; Ortenzi, F.; Pede, G.; Roncella, R.; Saletti, R. Experimental Analysis of an Electric Minibus with Small Battery and Fast Charge Policy. In Proceedings of the ESARS-ITEC 2016, Toulouse, France, 2–4 November 2016.

18. Alessandrini, A.; Cignini, F.; Ortenzi, F.; Pede, G.; Stam, D. Advantages of retrofitting old electric buses and minibuses. *Energy Procedia* **2017**, *126*, 995–1002. [CrossRef]

19. Alessandrini, A.; Barbieri, R.; Berzi, L.; Cignini, F.; Genovese, A.; Locorotondo, E.; Ortenzi, F.; Pierini, M.; Pugi, L. Design of a Hybrid Storage for Road Public Transportation Systems. In *Advances in Italian Mechanism Science. IFToMM ITALY 2018*; Mechanisms and Machine Science; Carbone, G., Gasparetto, A., Eds.; Springer: Cham, Switzerland, 2019; Volume 68.

Article

Fractional-Order Modeling and Parameter Identification for Ultracapacitors with a New Hybrid SOA Method

Jianhua Guo [1], Weilun Liu [1,*], Liang Chu [1] and Jingyuan Zhao [2]

[1] State Key Laboratory of Automotive Simulation and Control, Jilin University, Changchun 130022, China; xhhmail@jlu.edu.cn (J.G.); chuliang@jlu.edu.cn (L.C.)

[2] Auto Engineering Research Institute, BYD Auto Co., Ltd; Shenzhen 518118, China; zhao.jingyuan@byd.com

* Correspondence: wlliu17@mails.jlu.edu.cn

Received: 10 October 2019; Accepted: 7 November 2019; Published: 8 November 2019

Abstract: This paper deals with an ultracapacitor (UC) model and its identification procedure. To take UC's fractional characteristic into account, two constant phase elements (CPEs) are used to construct a model structure according to impedance spectrum analysis. The different behaviors of UC such as capacitance, resistance, and charge distribution dynamics are simulated by the corresponding part in the model. The resistance under different voltages is calculated through the voltage rebound method to explore its non-linear characteristics and create a look-up table. A nonlinear fractional model around an operation voltage is then deduced by applying the resistance table. This time identification is carried by a proposed hybrid optimization algorithm: Nelder-Mead seeker algorithm (NMSA), which embeds the Nelder–Mead Simplex (NMS) method into the seeker optimization algorithm (SOA). Its time behavior has been compared with the linear fractional model for charging and discharging current profiles at different levels.

Keywords: hybrid optimization; ultracapacitor; fractional-order model; time-domain analysis

---

1. Introduction

Ultracapacitors (UCs) are currently gaining attention rapidly as a new type of energy-storage component with a higher density than ordinary capacitance and longer cycle life than batteries [1,2]. However, its energy storage principle determines its capacity cannot temporarily surpass the battery [3]. Thence, it is always associated with batteries in the energy storage unit when it supplies energy to electric vehicles (EV) or hybrid electric vehicles (HEV). UC can meet vehicle startup and acceleration requirements due to its high power rate. Thanks to the short charging time of UC, the braking energy can be absorbed efficiently during decelerating. With a reasonable power distribution strategy in-vehicle, the application of UC is a good way to reduce battery load and further to improve the battery's cycle life.

In the early stage of UC modeling, the internal properties of UC are analyzed in many electrochemical models. In their model, Gouy and Chapman adopted the principle that ion distribution of the double-layer's outer element conformed to the laws of Bolzmann [4]. The mathematical model proposed in [5] described the effect of pore size on UC performance. On the contrary, the machine learning model describes the relationships between inputs and outputs without considering the internal structure. The subject [6–8] utilized ANN technology to predict the performance of carbon materials in supercapacitors. Among all the modeling approaches, the equivalent circuit model is the most commonly used. Through studying the slow charging behavior of UC, R.L. Spyker and R.M. Nelms used a classical equivalent circuit composed of capacitance in parallel with resistance and an equivalent series resistance [9]. To take into account the fast dynamics, a model consisting of three RC branches is proposed that reflected the internal charge distribution process within a specific voltage range [10]. Because of the widespread use of activated carbon with high surface area, some groups adopted the transmission line model (TLM) to simulate the frequency response of UC and discussed the contribution of the parameters to the impedance spectrum [11]. The complex determination of different elements and large numbers of parameters leads to inconvenience.

To simplify the complexity of the model, the fractional-order model has been employed. A simple UC model composed of a series resistor and a CPE is used to model the impedance in the frequency range in [12]. The capacitance-voltage dependence is taken into account in [13] by implanting a voltage-dependent capacitance in one of the three branches and a fitting formula with an undetermined coefficient is used. N. Bertrand et al. [14] used a set of linear systems to deduce a global nonlinear model.

In a process of model identification, there are traditional methods such as the least square methods, the gradient-based methods, and the maximum likelihood methods. They can be suitable for nonlinear systems after some efforts made by [15]. However, the recognition process of all the above methods needs to derive the objective function, which is difficult to implement in fractional-order systems. The derivative-free algorithms, such as the simplex search method [16] and Nelder–Mead simplex (NMS) algorithm, [17] are successfully applied to the parameter tuning. The limitations of these techniques are the natural local concern and the sensitivity to the initial solution of the algorithm. Fractional order system identification problems have been carried out using global heuristic techniques such as particle swarm optimization (PSO) [18], genetic algorithm (GA) [19], and artificial bee colony algorithm (ABC) [20]. Compared to the local search algorithm, they have a global performance nature and a greater possibility to converge to the global optimal point. Nevertheless, they often oscillate in the vicinity of the optimal point or miss the optimal area due to strong randomness.

The seeker optimization algorithm (SOA) is also a heuristic algorithm, which was invented by Dai [21]. It explores the experience of human search behavior through the analysis of the human random search process. By exchanging experience, seekers determine their direction and step size, then update their location, and, finally, find the optimal point. Since it is proposed, it has been used by researchers in different engineering applications like system modeling and optimal control [22,23]. To overcome the individual limits of SOA, a hybridization of SOA and NM algorithm is used in this paper, which simultaneously retains the benefits of both techniques while discarding their drawbacks.

The proposed modeling approach leads to a model that has the following characteristics:

- The model structure is connected to UC dynamic characteristics;
- The using of fractional order reduces the number of parameters;
- The resistance voltage dependence has been taken into account;
- The hybrid optimization algorithm improves recognition accuracy.

2. Fractional-Order Calculus

Fractional-order calculus (FOC) generalizes the integrals and derivatives from integer-order to arbitrary real order. This theory goes back to a note written by Leibniz, in which the meaning of the half-derivative of the function x was discussed. Some approaches to generalizations of the notion of calculus can be found in [24]. A great deal of work has been done to propose the models of linear time-invariant system and nonlinear system for analyzing the stability, time, frequency domain response, and other properties of the fractional-order model [25,26]. FOC provides an excellent instrument for the description of stochastic memory phenomena such as electrochemical phenomena, chaos, and viscoelasticity. Consider a function $y = f(t)$, which is integrable and has at least $m(m \epsilon N)$ continuous derivatives. The universal expression of integer-order derivatives and integrals is as follows:

$$_aD_t^p f(t) = \lim_{h \to 0} h^{-p} \sum_{r=0}^{n} (-1)^r \binom{p}{r} f(t - rh), \tag{1}$$

$$\left(\begin{array}{c} p \\ r \end{array} \right) = \frac{p(p-1)(p-2)\cdots(p-r+1)}{r!} \tag{2}$$

$\left(\begin{array}{c} p \\ r \end{array} \right)$ is the usual notation for the binomial coefficient, $n = [(t-a)/h]$. For $p = m(m > 0)$, $_aD_t^p = \frac{d^m t}{dt^m}$ represents the derivative of order m. On the contrary, $_aD_t^p$ denotes m fold integral if $p = -m(m > 0)$. The unification of integer-order derivatives and integrals is the basis for allowing p in (1) to be an arbitrary real number or even a complex number. In this section, three main definitions of fractional order integrals and derivatives are briefly introduced.

- *Grünwald–Letnikov Definition:*

$$_aD_t^\alpha f(t) = \lim_{h \to 0} h^{-\alpha} \sum_{r=0}^{n} (-1)^r \left(\begin{array}{c} \alpha \\ r \end{array} \right) f(t-rh), (m-1 < \alpha < m) \tag{3}$$

$$\left(\begin{array}{c} \alpha \\ r \end{array} \right) = \frac{1}{(\alpha+1)B(\alpha-r+1, r+1)} \tag{4}$$

- *Riemann–Liouville Definition:*

$$_aD_t^\alpha f(t) = \frac{1}{\Gamma(m-\alpha)} \frac{d^m}{dt} \int_a^t (t-\tau)^{m-\alpha-1} f(\tau)d\tau, (m-1 < \alpha < m) \tag{5}$$

- *Caputo definition:*

$$_aD_t^\alpha f(t) = \frac{1}{\Gamma(m-\alpha)} \int_a^t (t-\tau)^{m-\alpha-1} f^{(m)}(\tau)d\tau, (m-1 < \alpha < m) \tag{6}$$

Grünwald–Letnikov generalizes integer-order calculus directly and requires that $f(t)$ can be differentiated $m = [\alpha] + 1$ times ($\alpha \in R$). However, a limit operation is not convenient. Rieman–Liouville provides an excellent opportunity to weaken the requirement for the differentiability of $f(t)$ through m-α-1 times integral before m-order derivative. Another definition was proposed by M.Caputo that exchanges the sequence of derivative and integral. The main advantage of the Caputo approach is permitting the initial conditions for fractional differential equations to take on the same form as for integer-order differential equations, which makes them easy to interpret. Another difference between Caputo's definition and the other two definitions can be shown in the derivative of a constant C. The Caputo derivative of a constant is 0, whereas the other two bring a nonzero value as Equation (7), so the Caputo definition is used in this paper.

$$_aD_t^\alpha C = \frac{Ct^{-\alpha}}{\Gamma(1-\alpha)} \tag{7}$$

3. A New Hybrid Algorithm: NMSA

3.1. Seeker Optimization Algorithm (SOA)

SOA is a new swarm intelligence optimization algorithm, which has better optimization quality and robustness than GA and PSO [21]. It resolves optimization problems by means of map search. n seekers search for the lowest position on the D-dimension map if there are D optimized variables. The lowest position stands for the set of optimal parameters and the height of a point is calculated by the objective function. Their journey lasts for g days, to begin with, they are sent to the initial position randomly on the first day. On each of the following days, they need to record the lowest place they have ever been and the corresponding altitude. Meanwhile, they share their locations and altitudes with the information network, which can be read by everyone and the best location of the past few days can be selected. Afterward, according to several reference locations, each seeker determines the direction and distance to go tomorrow, following a certain rule, and start their search the next day. They move, communicate, and make plans at the end of the day journey until their whole journey is finished and a position closest to the bottom is found.

On kth day, ith seeker $p_i(k)$ has a tendency to move towards the personal best position $p_{ibest}(k)$ and the best position of population $p_{zbest}(k)$. Meanwhile, he tends to inherit the moving direction $d_i(k-1)$ of the yesterday if $f(p_i(k)) < f(p_i(k-1))$ and to reverse it if not. Another reference direction might also exist, which is represented by $d_{iother}(k)$. Those reference directions are marked in Figure 1. Using horizontal as an example, after the number of reference directions toward the positive direction and those toward the negative direction are counted, respectively, draw the probability distribution semicircle. Then ith seeker determines his moving direction $d_i(k)$ by generating a random angle. If the angle is as shown in Figure 1, the ith seeker moves positively in horizontal.

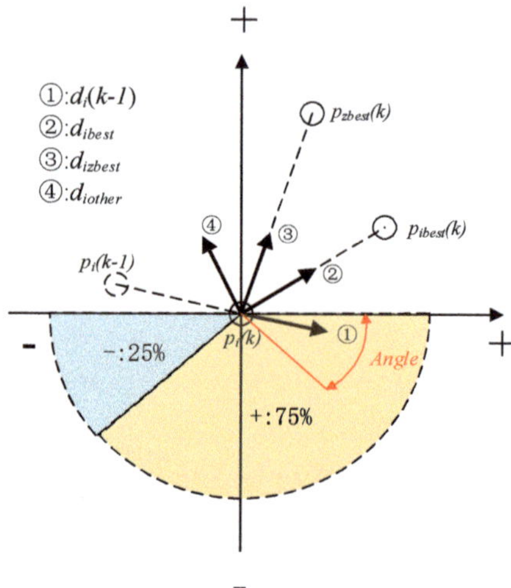

Figure 1. Determine the search direction.

The search step length $l_i(k)$ is calculated by Gaussian membership function in Equation (8). μ_i is a decreasing function of ith seeker's fitness ranking I_i. I_i means that $f(p_i(k))$ is the I_ith lowest in all $f(p(k))$. σ_i is negatively correlated with k. μ_i allows the seeker near the optimal point to search for more details and σ_i makes seekers search more carefully at the later stage of the search. Figure 2 is an intuitive expression for determining the step length. '→' represents the tendency to lessen step length.

$$\mu(l_i) = e^{-\frac{l_i^2}{2\sigma_i^2}} \tag{8}$$

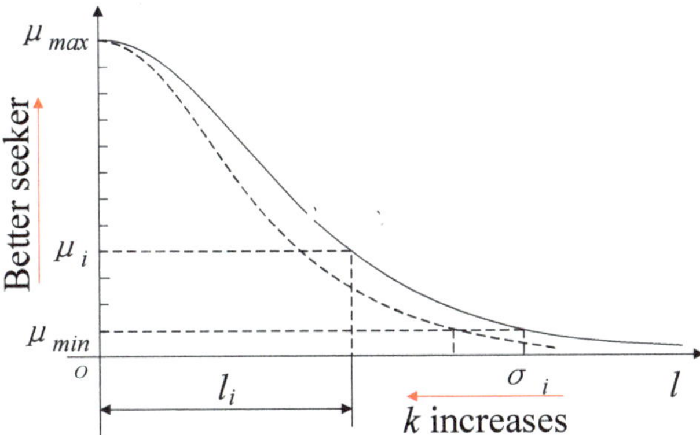

Figure 2. Determine the search step length.

3.2. Nelder–Mead Simplex Algorithm

As a derivative-free method, the NM algorithm is used to construct a simplex to solve unconstrained minimization problems. To minimize a mathematical function f that has n real parameters, the simplex is designed with $n + 1$ vertexes and to replace the worst vertex with a better one each iteration via reflection, expansion, contraction, or reduction.

After the initial simplex is determined, the function f is evaluated at each of the vertices, which are later renamed according to its value and then comes the relationship in Equation (9). The basis points and the candidate points that will be used later are listed in Table 1. And they are elected pursuant to the rule shown in Figure 3. The NM algorithm is the first step towards finding an effective substitute for X_w along the line segment joining X_c and X_w as seen in Figure 4. If the first step fails, NM no longer explores that line segment but starts to draw closer to X_b by shrinking. Figure 5 shows the evolution of simplex when NM is used to find the optimal point of a 2-dimensional objective function.

$$f(X_1) \leq f(X_2) \leq \cdots \leq f(X_{n+1}) \tag{9}$$

Table 1. Basis points and the candidate points of Nelder–Mead (NM).

Basis		Candidate	
Best vertex	$X_b = X_1$	Reflection point	$X_r = X_c + \alpha(X_c - X_w)$
Second worst vertex	$X_{sw} = X_n$	Expansion point	$X_e = X_c + \beta(X_r - X_c)$
Worst vertex	$X_w = X_{n+1}$	Outer contraction point	$X_{oc} = X_c + \gamma(X_r - X_c)$
Centroid	$X_c = \frac{1}{n}\sum_{i=1}^{n} X_i$	Inner contraction point	$X_{ic} = X_c + \gamma(X_w - X_c)$
		Shrinkage points	$X_s(i) = X_b + \delta(X_i - X_b)$ $i = 2, 3, \ldots, n+1$

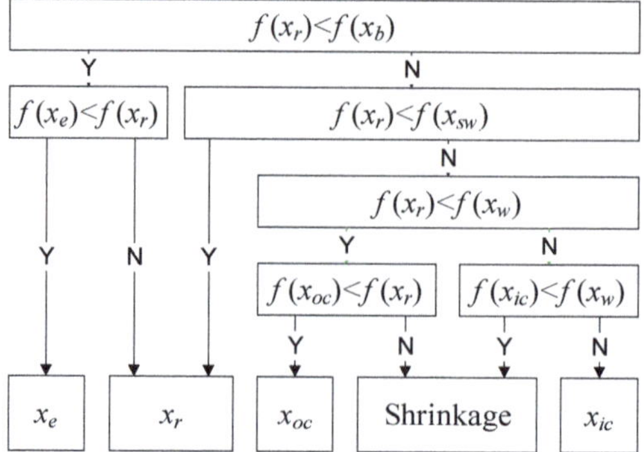

Figure 3. NM evolutionary process.

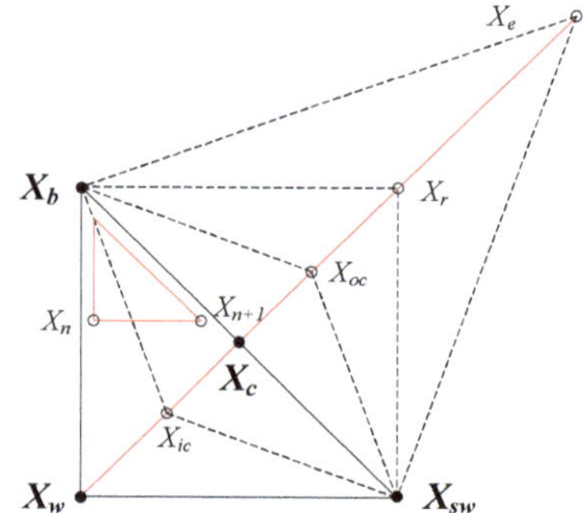

Figure 4. NM algorithm.

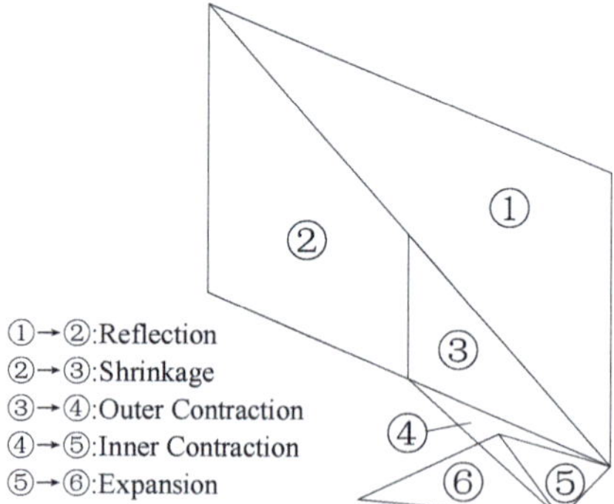

Figure 5. A case of NM algorithm.

3.3. New Hybrid SOA: NMSA

The random factors in the seeker's direction and step length bring SOA good global exploration ability. Nevertheless, it is also the randomness that causes its poor local exploration ability. SOA's later search is always wasted. Besides, when dealing with a multi-extremum problem, SOA may converge to a wrong extreme point because of insufficient exploration to the main extreme point. As a beam search method, the NM simplex method has opposite capability compared with SOA. However, it also produces incorrect outputs due to the quality of the initial simplex. In this paper, a synthetic algorithm of NM and SOA is proposed to make their respective advantages complementary to each other. In every day's searching, the top $n + 1$ seekers of SOA do the NM transformation and iterate once. After that, they begin to share their positions and to make the next day's plan. A diagram of the proposed method NMSA (The hybrid algorithm based on the seeker algorithm and the Nelder–Mead simplex) is given in Figure 6.

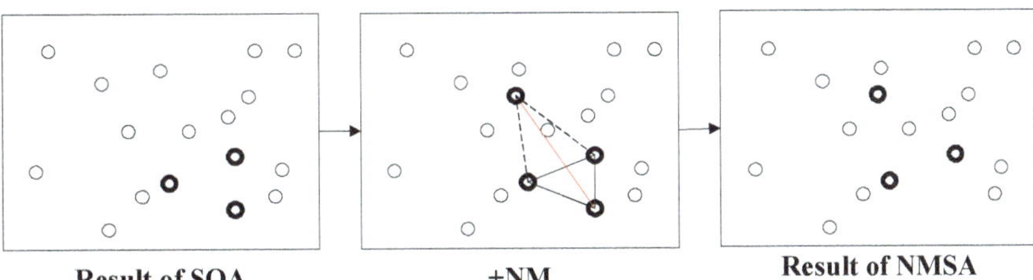

Figure 6. NMSA algorithm.

To evaluate the performance of the optimization strategy, a set of artificial test functions has been proposed and used as benchmarks. Typically, they can be classified in the term of modality. If there is only one peak in the function landscape, the function is said to be unimodal (e.g., sphere function, booth function, etc.), which is often used for convergence velocity comparison [27]. On the contrary, the function is a multimodal function (e.g., Rastrigin function, Ackley function, Schwefel function, etc.). It turns out to be an extremely important performance test for global optimization because it creates difficulties in finding a global minimum. In such a landscape, the search process can direct the search away from the true optimal solutions and trapped in one of the local minima.

For our proposed offline global optimization algorithm NMSA, its reliability and precision should be tested. Rastrigin function, one of the multimodel functions, is based on De Jong's function with the addition of cosine modulation to produce many local minima. In addition, the small fitness differences in the topology increase the complexity of the problem [28]. Therefore, consider the Rastrigin function as the test function of NMSA. The function is given by the following formula and visualized by Figure 7.

$$f(x, y) = x^2 + y^2 - 10\cos(2\pi x) - 10\cos(2\pi y) + 20 \tag{10}$$

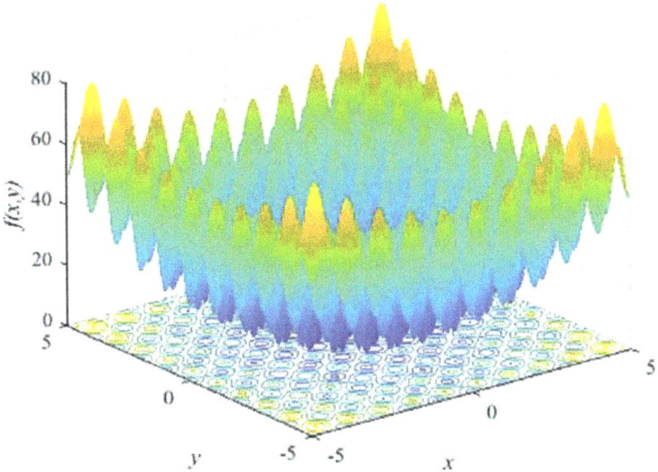

Figure 7. Visualization of Rastrigin function.

Rastrigin function has several minima. The global one is at (0, 0). This minimum gives a function value $f = 0$. The nearest other minima are, respectively, at (0, 1), (0, −1), (1, 0), and (−1, 0), and they all give a value $f = 1$. As shown in Figure 8, based on the respective function value obtained by NMSA and SOA after 10 runs, it can be noted that the NMSA algorithm leads successfully to the global minimum compared with the original SOA. With SOA, nearly half of the optimization processes converge to secondary minima and the remainder need at least 90 iterations to converge to the global minimum. NMSA needs less than 60 iterations to get the correct minimum with no wrong judgments in 10 runs. That is because NM offsets the SOA's weakness in local exploration and improves the directionality in NMSA. In general, NMSA has the following characteristics: (1) Its global and local exploration abilities are well balanced for finding the global minimum; (2) It has combined the ergodicity and directionality effectively for a faster rate of convergence. In Table 2, the search results after 100 iterations are presented in two numerical forms: the coordinates of the searched approximate optimal solution and the corresponding fitness value. The former shows that NMSA has advantages over SOA in finding global optimization. As for the latter, the fitness value is numerically equal to the fitness error because the true optimal fitness value in this example is 0. After comparing this value, we can see that the search accuracy of NMSA is about two orders of magnitude higher than that of SOA.

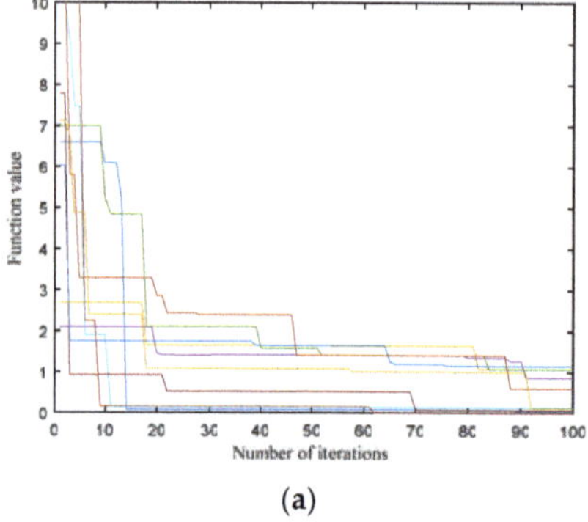

(a)

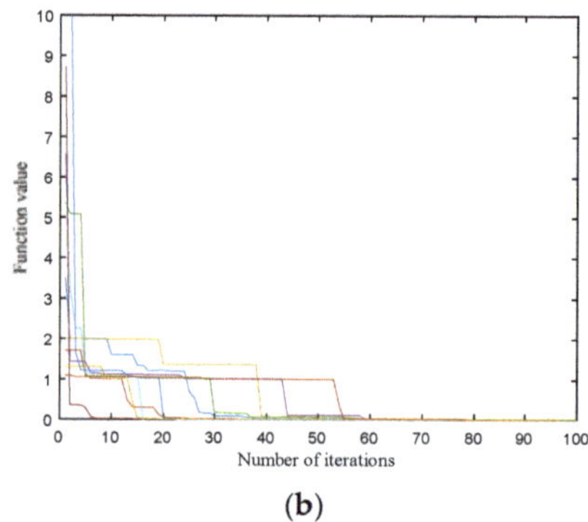

(b)

Figure 8. The Fitness curve of the seeker optimization algorithm (SOA) and NMSA; (**a**) is the fitness curve of SOA; (**b**) is the fitness curve of NMSA.

Table 2. Results of SOA and NMSA.

$(x,y)_{searched\ by\ SOA}$	$f(x,y)$	$(x,y)_{searched\ by\ NMSA}$	$f(x,y)$
(+8.8432E−3, −1.1794E−2)	+4.3093E−2	(+1.1729E−3, +6.7883E−3)	+9.4136E−3
(−2.0045E−3, −2.9132E−3)	+2.4808E−3	(−4.1521E−3, +3.3201E−5)	+3.4204E−3
(−1.0142E00, −6.2758E−3)	+1.0763E00	(−4.3177E−4, −1.2294E−3)	+3.3684E−4
(+6.6242E−2, −5.0537E−3)	+8.6317E−1	(+2.7027E−3, −7.0799E−3)	+1.1392E−2
(+1.0154E00, −6.6096E−3)	+1.0867E00	(−9.5493E−3, +3.2802E−3)	+2.0220E−2
(+2.1976E−2, −1.4773E−2)	+1.3893E−1	(+2.2814E−3, +6.8896E−4)	+1.1267E−3
(−1.7572E−2, +8.4259E−3)	+7.5280E−2	(−2.1941E−3, −2.5050E−2)	+4.4583E−3
(−9.7707E−1, −2.2125E−2)	+1.1553E00	(−3.0697E−4, +2.6148E−3)	+1.3751E−3
(−2.0418E−2, −5.1174E−2)	+5.9771E−1	(−5.7813E−4, −1.8963E−3)	+7.7976E−4
(+2.3927E−2, +2.9333E−3)	+1.1507E−1	(−9.8208E−4, +1.1038E−4)	+1.9376E−4

4. Modeling and Identification of UC

4.1. Fractional Order Modeling of UC

UC, also known as electrochemical double layer capacitor, stores the electrical energy by ion absorption in the double-layer or pseudo-capacitive between electrode and electrolyte. A double layer in the order of 0.3–0.8 nm gradually forms during charging or discharging, which leads to a larger capacitance and higher energy density than ordinary capacitors. In addition, carbon with pores is widely used as the electrode of UC because it provides a larger surface area. Because of that energy storage principle and the use of carbon electrodes, UC performance is affected by distributed surface reactivity, inhomogeneity, and roughness, which can be better characterized by CPE according to the electrochemical impedance spectroscopy (EIS). The impedance curve will deflect downward to a squashed semi-circle if the capacitor parallel with the resistor is replaced by CPE. The CPE can also model the imperfect capacitor behavior of a double-layer capacitor; therefore, this paper proposed a fractional model with 2 CPEs as shown in Figure 9.

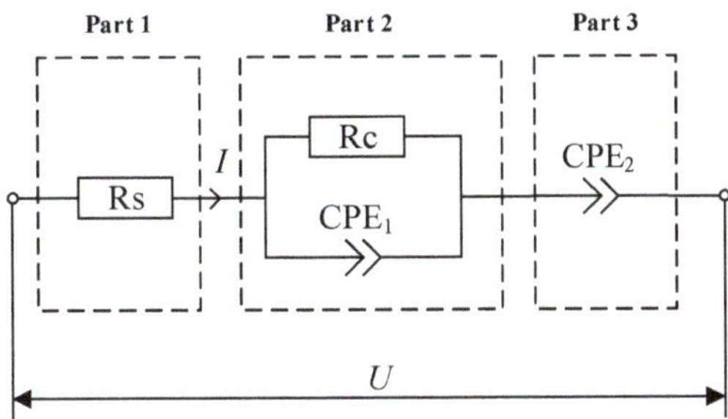

Figure 9. Ultracapacitor (UC) fractional model.

The impedance of CPEs can be expressed as follows:

$$Z_{CPE_1}(j\omega) = \frac{1}{C_1(j\omega)^\alpha} \tag{11}$$

$$Z_{CPE_2}(j\omega) = \frac{1}{C_2(j\omega)^\beta} \tag{12}$$

where C_1 and C_2 are the capacitance coefficients, $\alpha(0 \leq \alpha \leq 1)$ and $\beta(0 \leq \beta \leq 1)$ are the fractional orders of CPE. According to the equivalent circuit structure, the transfer function of the fractional model can be delineated by the following:

$$G(s) = \frac{U(s)}{I(s)} = R_s + \frac{R_c}{1 + C_1 R_c s^\alpha} + \frac{1}{C_2 s^\beta} \tag{13}$$

where R_s represents the series resistance and R_s is particularly small for UC. Part 2 which consists of CPE_1 and R_c serves to capture the charge diffusion caused by specifically adsorbed ions. CPE_2 models the behavior of the double layer. To simplify the solving process, the inverse Laplace transformation in Equation (13) can be divided into three parts due to its linear nature:

$$U(t) = U_1(t) + U_2(t) + U_3(t) \tag{14}$$

$$\frac{U_1(t)}{R_s} = I(t) \tag{15}$$

$$\frac{U_2(t)}{R_c} + C_1 D_t^\alpha U_2(t) = I(t) \tag{16}$$

$$C_2 D_t^\beta U_3(t) = I(t) \tag{17}$$

With the method proposed in [29] by Zhao C.N. and Xue D.Y., the solution to the fractional-order differential equation

$$a_1 D^{\gamma_1} y(t) + a_2 D^{\gamma_2} y(t) + \cdots + a_n D^{\gamma_n} y(t) = u(t) \tag{18}$$

can be written as

$$y(t) = \frac{u(t) - \sum_{i=1}^{n}\left(\frac{a_i}{h^{\gamma_i}} \sum_{j=1}^{[(t-t_0)/h]} \omega_j^{(\gamma_i)} y_{t-jh} \right)}{\sum_{i=1}^{n} \frac{a_i}{h^{\gamma_i}}} \tag{19}$$

$$\omega_0^{(\gamma_i)} = 1, \omega_j^{(\gamma_i)} = \left(1 - \frac{\gamma_i + 1}{j}\right)\omega_{j-1}^{\gamma_i}, \ j = 1, 2, \cdots \tag{20}$$

For part 1: $U_1(t) = R_s \cdot I(t)$;
For part 2: $u(t) = I(t), y(t) = U_2(t), a = \begin{bmatrix} C_1 & \frac{1}{R_c} \end{bmatrix}, \gamma = \begin{bmatrix} \alpha & 0 \end{bmatrix}$;
For part 3: $u(t) = I(t), y(t) = U_3(t), a = C_2, \gamma = \beta$.

4.2. Time Domain Response Analysis of UC

After the current is set to zero, the terminal voltage of UC varies with the time. This phenomenon is called voltage rebound and can be used to calculate the effective resistance of UC with the formula $R_t = U_t/I$ [30]. The response of Nesscap to a single pulse is drawn in Figure 10. When current is suddenly set to zero, $t = 0$ and the resistance at this point $R_0 = U_0/I$. As t increases, R_t increases. If parameters are assigned values: $R_s = 1.537 \text{ m}\Omega$, $R_c = 5.393 \text{ m}\Omega$, $C_1 = 7501 \text{ F}$, $C_1 = 2918 \text{ F}$, $\alpha = 0.2699$, $\beta = 0.9663$, the model response fits well with the real response. Decompose the model response into three parts, as shown in Figure 10, according to the previous section. U_1 imitates the immediate resistance rebound at $t = 0$ and equals to U_0, so $R_s = R_0$. It can be seen both U_2 and U_3 contribute to the response for $t > 0$ and $t < 0$. U_2 mainly represents the obvious inertia properties under the sudden current change while U_3 mainly represents the capacitance characteristics. The existence of α and β provides a more flexible approach to receive a more accurate model.

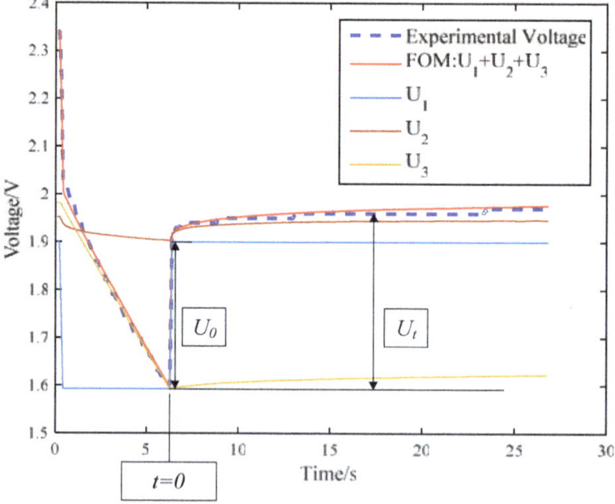

Figure 10. Single-pulse response of UC.

The resistance of different voltages at 200 A is calculated by using the same method and its variation with voltage is shown in Figure 11. Charging R_s keeps steady in the charging process. In discharging experiment, R_s increases sharply. In the later parameter identification, R_s is determined by the look-up table method.

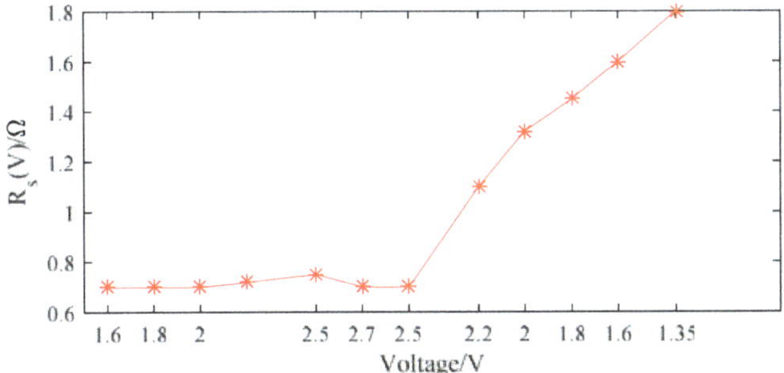

Figure 11. Resistance variation at 200 A.

4.3. Parameter Estimation in Time Domain

A Nesscap UC with the nominal capacitance of 3000 F, steady-state resistance of 0.4 mΩ, power density of 512 W/kg, and energy density of 4.4 Wh/kg is used to carry on charge and discharge experiments at different currents. The rectangular current with an amplitude value of 200 A is the input during charging. When R_s can be determined using the look-up table, the remaining five parameters are identified by NMSA. Its objective function is defined as Equation (21), which can well evaluate the ability of the model to describe the real capacitance behavior. $U(t)$ is the measured voltage and $\hat{U}(t)$ is the derived result from identified model. Sampling time $t = 0.1$ s. The key parameters of NMSA are as follows: population size $S = 20$, number of searching days $g = 100$, dimension of map is 5, reflection factor $\alpha = 1$, expansion factor $\beta = 2$, contraction factor $\gamma = 0.5$, shrinkage factor $\delta = 0.5$.

$$fitness = \sqrt{\frac{\sum_{j=1}^{[(t_{end}-t_0)/h]}\left(U(j) - \hat{U}(j)\right)}{[(t_{end} - t_0)/h]}} \tag{21}$$

5. Results and Discussion

After performing the optimization procedure, the result is $R_c = 14.31$ mΩ, $C_1 = 7021$ F, $C_2 = 3799$ F, $\alpha = 0.7944$, $\beta = 0.9994$ for charging. The validation test has been done with charging and discharging currents at levels of 300 A and 400 A. The UC voltage is constrained to vary in the voltage range [1.35 V, 2.7 V]. The results are presented in Figure 12. Compared to the fitness results obtained by NMSA without considering the voltage dependence of R_s, great improvement was achieved under a discharge condition. The error between Nesscap and the nonlinear model is within 0.02 V for charging and 0.05 V for discharging. The result shows that the model gives an accurate voltage response for 300 A and 400 A current level. Though there is a slight difference between the charging and discharging, which may be caused by changes in other parameters, it can be neglected. The proposed model simulates UC behavior with good accuracy in the time domain.

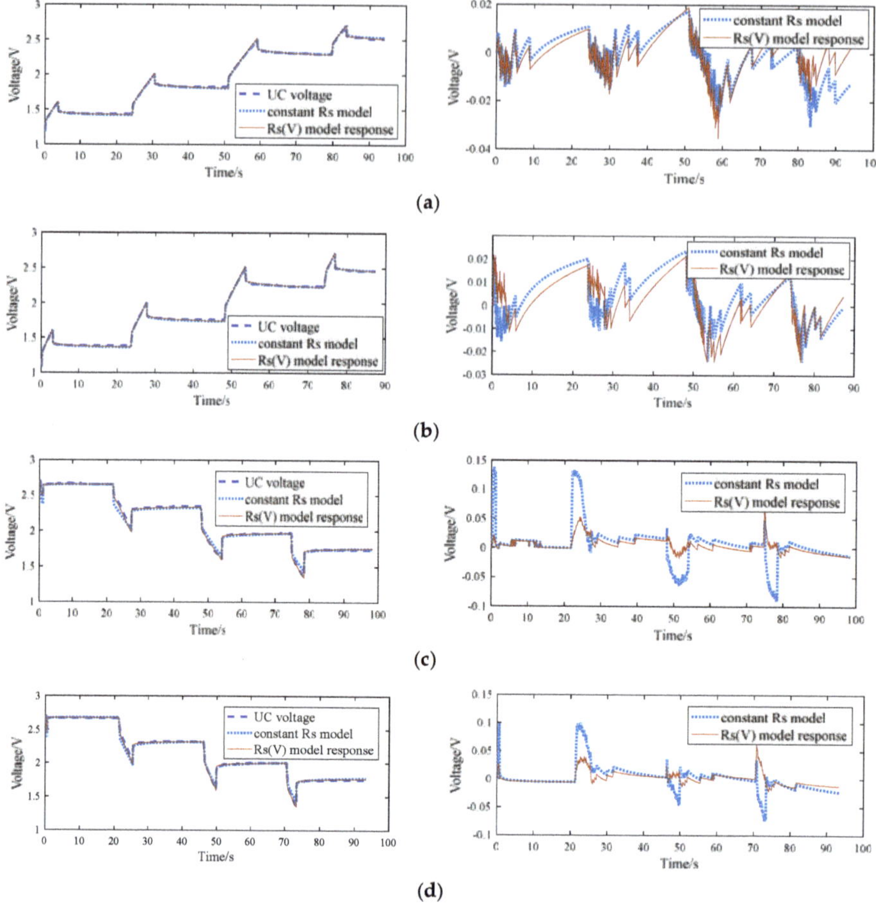

Figure 12. Comparison between model voltage response and UC voltage (The left side is the response curve and the right side it the response error compared with UC voltage) (**a**) Validation test with 300 A charging profile; (**b**) Validation test with 400 A charging profile; (**c**) Validation test with 300 A discharging profile; (**d**) Validation test with 400 A discharging profile.

6. Conclusions

In this paper, a fractional model of UC is applied. The influence of two fractional parameters $\alpha(0 \leq \alpha \leq 1)$ and $\beta(0 \leq \beta \leq 1)$ on the UC model's time domain is analyzed. α represents the inertia properties in UC, β represents the undesirable components in UC. The voltage dependence of R_s is discussed and a look-up table is set up to add nonlinear characteristics to the model using the voltage rebound method. The resistance of UC varies greatly during discharging. A new hybrid optimization algorithm NMSA is proposed in this paper and has been proven to have sufficient global search capability and accuracy. After applying 300 A and 400 A level current to the identified model, it can be proved that the current magnitude and status of charging or discharging have almost no influence or slight influence on the UCs behavior in the range [200 A, 400 A].

Author Contributions: Conceptualization, W.L. and J.Z.; methodology, W.L.; software, W.L.; validation, W.L.; formal analysis, J.Z.; investigation, W.L. and J.Z.; resources, J.Z.; data curation, J.Z.; writing—original draft preparation, W.L.; writing—review and editing J.G. and J.Z.; visualization, W.L.; supervision, J.G.; project administration, J.G.; funding acquisition, L.C.

Funding: The APC was funded by the Key Science and Technology Development Project: Research on the driving cycle development and the control strategy for PHEV based on the Intelligent Transportation System (Project No. 20170204085GX).

Conflicts of Interest: The authors declare no conflict of interest.

References

1. Kurra, N.; Jiang, Q.; Alshareef, H.N. A general strategy for the fabrication of high performance microsupercapacitors. *Nano Energy* **2015**, *16*, 1–9. [CrossRef]
2. Ladrón de Guevara, A.; Boscá, A.; Pedrós, J.; Climent-Pascual, E.; de Andrés, A.; Calle, F.; Martínez, J. Reduced Graphene oxide/polyaniline electrochemical supercapacitors fabricated by laser. *Appl. Surf. Sci.* **2019**, *467*, 691–697. [CrossRef]
3. Conway, B.E. *Electrochemical Supercapacitors: Scientific Fundamentals and Technological Applications*, 1st ed.; Springer Science & Business Media: New York, NY, USA, 2013; pp. 11–31.
4. Oldham, K.B. A Gouy–Chapman–Stern model of the double layer at a (metal)/(ionic liquid) interface. *J. Electroanal. Chem.* **2008**, *613*, 131–138. [CrossRef]
5. Staser, J.A.; Weidner, J.W. Mathematical modeling of hybrid asymmetric electrochemical capacitors. *J. Electrochem. Soc.* **2014**, *161*, E3267–E3275. [CrossRef]
6. Eddahech, A.; Ayadi, M.; Briat, O.; Vinassa, J.-M. Multilevel neural-network model for supercapacitor module in automotive applications. In Proceedings of the 4th International Conference on Power Engineering, Energy and Electrical Drives, Istanbul, Turkey, 13–17 May 2013; pp. 1460–1465.
7. Zhu, S.; Li, J.; Ma, L.; He, C.; Liu, E.; He, F.; Shi, C.; Zhao, N. Artificial neural network enabled capacitance prediction for carbon-based supercapacitors. *Mater. Lett.* **2018**, *233*, 294–297. [CrossRef]
8. Ma, H.; Wang, B.; Liu, K.R. Distributed Signal Compressive Quantization and Parallel Interference Cancellation for Cloud Radio Access Network. *IEEE Trans. Commun.* **2018**, *66*, 4186–4198. [CrossRef]
9. Spyker, R.L.; Nelms, R.M. Classical equivalent circuit parameters for a double-layer capacitor. *IEEE Trans. Aerosp. Electron. Syst.* **2000**, *36*, 829–836. [CrossRef]
10. Zubieta, L.; Bonert, R. Characterization of double-layer capacitors for power electronics applications. *IEEE Trans. Ind. Applicat.* **2000**, *36*, 199–205. [CrossRef]
11. Itagaki, M.; Suzuki, S.; Shitanda, I.; Watanabe, K.; Nakazawa, H. Impedance analysis on electric double layer capacitor with transmission line model. *J. Power Sources* **2007**, *164*, 415–424. [CrossRef]
12. Freeborn, T.J.; Maundy, B.; Elwakil, A.S. Fractional-order models of supercapacitors, batteries and fuel cells: A survey. *Mater. Renew. Sustain. Energy* **2015**, *4*, 9. [CrossRef]
13. Bertrand, N.; Sabatier, J.; Briat, O.; Vinassa, J.-M. Embedded fractional nonlinear supercapacitor model and its parametric estimation method. *IEEE Trans. Ind. Electron.* **2010**, *57*, 3991–4000.
14. Li, M.; Liu, X. The least squares based iterative algorithms for parameter estimation of a bilinear system with autoregressive noise using the data filtering technique. *Signal Process.* **2018**, *147*, 23–34. [CrossRef]
15. Ding, F.; Chen, H.; Xu, L.; Dai, J.; Li, Q.; Hayat, T. A hierarchical least squares identification algorithm for Hammerstein nonlinear systems using the key term separation. *J. Frankl. Inst.* **2018**, *355*, 3737–3752. [CrossRef]
16. Spendley, W.; Hext, G.R.; Himsworth, F.R. Sequential application of simplex designs in optimisation and evolutionary operation. *Technometrics* **1962**, *4*, 441–461. [CrossRef]
17. Nazareth, L.; Tseng, P. Gilding the lily: A variant of the Nelder-Mead algorithm based on golden-section search. *Comput. Optim. Appl.* **2002**, *22*, 133–144. [CrossRef]
18. Sun, J.; Zhao, J.; Wu, X.; Fang, W.; Cai, Y.; Xu, W. Parameter estimation for chaotic systems with a drift particle swarm optimization method. *Phys. Lett. A* **2010**, *374*, 2816–2822. [CrossRef]
19. Zhang, L.; Hu, X.; Wang, Z.; Sun, F.; Dorrell, D.G. Fractional-order modeling and State-of-Charge estimation for ultracapacitors. *J. Power Sources* **2016**, *314*, 28–34. [CrossRef]
20. Li, X.; Yin, M. Parameter estimation for chaotic systems by hybrid differential evolution algorithm and artificial bee colony algorithm. *Nonlinear Dyn.* **2014**, *77*, 61–71. [CrossRef]
21. Dai, C.; Chen, W.; Cheng, Z.; Li, Q.; Jiang, Z.; Jia, J. Seeker optimization algorithm for global optimization: A case study on optimal modelling of proton exchange membrane fuel cell (PEMFC). *J. Electr. Power* **2011**, *33*, 369–376. [CrossRef]

22. Ketabi, A.; Sheibani, M.R.; Nosratabadi, S.M. Power quality meters placement using seeker optimization algorithm for harmonic state estimation. *J. Electr. Power* **2012**, *43*, 141–149. [CrossRef]
23. Kumar, M.R.; Deepak, V.; Ghosh, S. Fractional-order controller design in frequency domain using an improved nonlinear adaptive seeker optimization algorithm. *Turk. J. Electr. Eng. Comput.* **2017**, *25*, 4299–4310. [CrossRef]
24. Ross, B. A brief history and exposition of the fundamental theory of fractional calculus. In *Fractional Calculus and Its Applications*, 1st ed.; Ross, B., Ed.; Springer: Berlin, German, 1975; Volume 457, pp. 1–36.
25. Monje, C.A.; Chen, Y.; Vinagre, B.M.; Xue, D.; Feliu-Batlle, V. *Fractional-Order Systems and Controls: Fundamentals and Applications*, 1st ed.; Springer Science & Business Media: London, UK, 2010; pp. 9–34.
26. Dzieliński, A.; Sierociuk, D.; Sarwas, G. Some applications of fractional order calculus. *Bull. Pol. Acad. Sci. Tech. Sci.* **2010**, *58*, 583–592. [CrossRef]
27. Mühlenbein, H.; Schomisch, M.; Born, J. The parallel genetic algorithm as function optimizer. *Parallel Comput.* **1991**, *17*, 619–632. [CrossRef]
28. Thomas, B. *Evolutionary Algorithms in Theory and Practice: Evolution Strategies, Evolutionary Programming, Genetic Algorithms*, 1st ed.; Oxford University Press: New York, NY, USA, 1996; pp. 137–159.
29. Xue, D.; Bai, L. Numerical algorithms for Caputo fractional-order differential equations. *Int. J. Control* **2017**, *90*, 1201–1211. [CrossRef]
30. Zhao, J.; Gao, Y.; Burke, A.F. Performance testing of supercapacitors: Important issues and uncertainties. *J. Power Sources* **2017**, *363*, 327–340. [CrossRef]

Article

Memory Effect and Fractional Differential Dynamics in Planar Microsupercapacitors Based on Multiwalled Carbon Nanotube Arrays

Evgeny P. Kitsyuk [1], Renat T. Sibatov [2,*] and Vyacheslav V. Svetukhin [1]

[1] Scientific-Manufacturing Complex "Technological Centre", Moscow 124498, Russia; kitsyuk.e@gmail.com (E.P.K.); svetukhin@mail.ru (V.V.S.)

[2] Laboratory of Diffusion Processes, Ulyanovsk State University, Ulyanovsk 432017, Russia

* Correspondence: ren_sib@bk.ru

Received: 6 December 2019; Accepted: 27 December 2019; Published: 2 January 2020

Abstract: The development of portable electronic devices has greatly stimulated the need for miniaturized power sources. Planar supercapacitors are micro-scale electrochemical energy storage devices that can be integrated with other microelectronic devices on a chip. In this paper, we study the behavior of microsupercapacitors with in-plane interdigital electrodes of carbon nanotube array under sinusoidal excitation, step voltage input and sawlike voltage input. Considering the anomalous diffusion of ions in the array and interelectrode space, we propose a fractional-order equivalent circuit model that successfully describes the measured impedance spectra. We demonstrate that the response of the investigated micro-supercapacitors is linear and the system is time-invariant. The numerical inversion of the Laplace transforms for electric current response in an equivalent circuit with a given impedance leads to results consistent with potentiostatic measurements and cyclic voltammograms. The use of electrodes based on an ordered array of nanotubes reduces the role of nonlinear effects in the behavior of a supercapacitor. The effect of the disordering of nanotubes with increasing array height on supercapacitor impedance is considered in the framework of a distributed-order subdiffusion model.

Keywords: supercapacitor; anomalous diffusion; fractional derivative; equivalent circuit; distributed-order subdiffusion

1. Introduction

Planar micro-supercapacitors (PMSCs) are micro-scale electrochemical energy storage devices that can be embedded in a chip and integrated with other microelectronic devices. Different materials were suggested for electrodes of PMSC including graphene, nanowires, multiwalled carbon nanotubes (MWCNT), and conducting polymers (see review [1] and references therein). With an in-plane interdigital structure (Figure 1a,b), the PMSCs have a better performance than the conventional sandwich-like analogs of similar size [1]. Due to the spatial separation of the electrodes there is no need for a separator and organic binders. This simplifies the design and simulation of the device [2]. Interdigitized PMSCs possess excellent mechanical and electrical characteristics [3]. Due to design, the decreased ionic diffusion path between electrodes leads to small ion transport resistance, that results in higher power capability [3]. The high electrical conductivity of electrodes and their porosity can mitigate the ion diffusion resistance caused by the electrolyte viscosity [4].

An important problem is the development of reliable models for engineering calculations and diagnostics of PMSC. The simplest way is to extend the effective models of standard supercapacitors taking the geometry and specifics of PMSCs into account. One of the successful directions in phenomenological modeling of conventional supercapacitors is based on the use of fractional calculus [5–12]. Usually, these models are formulated in terms of equivalent circuits containing a constant phase element (CPE) sometimes called a capacitor of fractional order or a fractal element [13–15]. This element has an impedance with a frequency dependence proportional to $(j\omega)^{-\alpha}$. Bertrand, Sabatier et al., and Allagui et al. [8,16–18] have shown that fractional order models can effectively describe the dynamical behavior of electrochemical power sources and can be used to build SOC (state of charge) and/or SOH (state of health) estimators. Fractional differential models of supercapacitors were used to describe their operation in pulse power systems [19], to describe the impedance in a frequency range [5,7,20], to study temporal response of supercapacitors [11,21–23], to identify parameters from transient characteristics [24,25], to control supercapacitor systems using fractional state models [7], or fractional nonlinear systems [8,26].

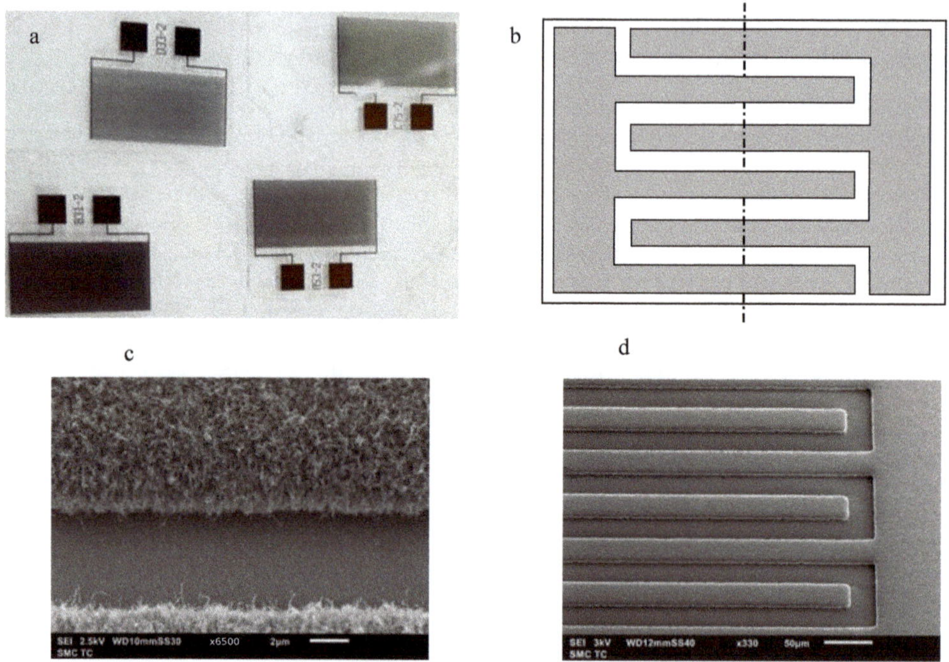

Figure 1. Planar supercapacitors studied in this work: photographs of several samples (**a**), planar geometry of PMSC (**b**), SEM images of multiwalled carbon nanotube (MWCNT) array in electrodes for different resolutions (**c,d**).

In this paper, for the first time, a fractional differential model for PMSC is proposed. We show that the response of the studied supercapacitors with carbon nanotube-based electrodes is linear and the system is time-invariant. The proposed model well describes the behavior of PMSC that is confirmed by various experimental methods. We assume that the use of electrodes with an array of rod-like nanostructures should reduce the role of nonlinear effects in PMSC behavior. The effect of weak disordering of CNT array for increased height is considered in the framework of a distributed-order subdiffusion model.

2. Carbon Nanotubes in Planar Microsupercapacitors

The studied PMSC samples were fabricated on silicon or borosilicate glass substrates by means of standard microelectronics technologies. In the case of silicon substrates, they were pre-oxidized to form a layer of high-quality dielectric. The oxide thickness was ~0.3 µm. Next, a layer of metal (titanium) was applied to the plates. The measurements were carried out under normal atmospheric conditions. During measurement, the samples were placed into the shielded volume to protect against electromagnetic interference. Contacting the electrodes on the chip was carried out using silver-plated clamps. Multiwalled carbon nanotubes were synthesized by catalytic PECVD process (see details in [27]) (Figure 1). The capacities of a supercapacitor with a solid electrolyte based on polyvinyl alcohol (PVA) were investigated. Phosphoric acid H_3PO_4 or lithium chloride LiCl were used as the ion-conducting agent.

The model developed here is applied to description of PMSC with electrodes composed of MWCNT forest. Two samples of PMSC are analyzed. They are labeled as D33 and A53, and differ in the width of stripes with MWCNTs: 30 μm for D33 and 50 μm for A53. Otherwise, they are the same, including the area. The measured voltammograms (Figure 2) for fabricated PMSCs demonstrate shapes indicating the absence of electrochemical reactions on electrode-electrolyte interface. These shapes are typical for supercapacitors, where electric double layer plays the main role in charging kinetics. As shown below, these samples are characterized by linear response uniform in time that make these devices very convenient for description and prediction of their behavior under various conditions.

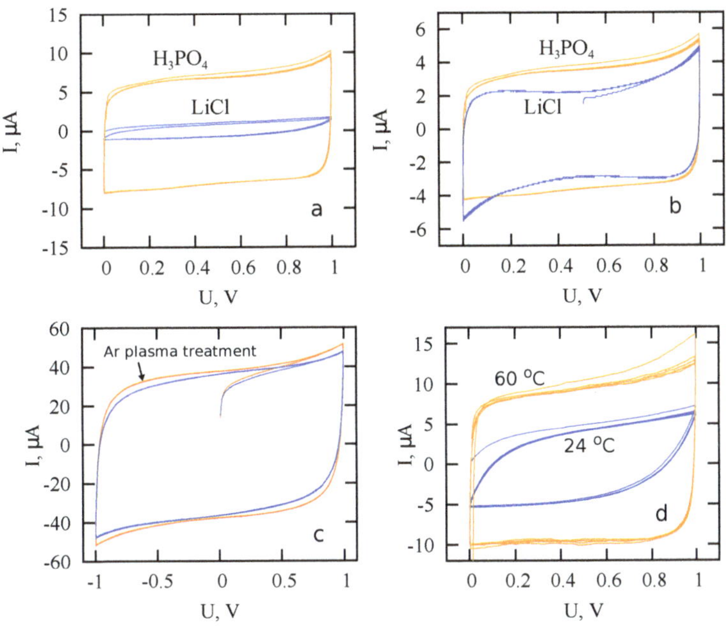

Figure 2. Cyclic voltammograms of planar micro-supercapacitor with solid electrolyte PVA + H_3PO_4 for scanning rates 50 mV/s (**a**) and 25 mV/s (**b**) as well as for the plasma treated sample (**c**) and curves measured at two different temperatures (**d**).

3. Linear Response and Anomalous Diffusion Model

In this section, we consider the linear response in terms of the integral equation describing a renewal generally non-Markovian process. By using the Mellin tranform of its kernel, it can be shown that a time-invariant dynamical process obeys a fractional differential equation of a complex order without additional assumptions [28]. The process is characterized by the order (spectral) distribution, so this representation incorporates both the standard Debye process and the non-Debye process family depending on choice of the spectral distribution. .

Generally speaking, the object under investigation in the linear response theory (LRT) is an abstract system with two points: input and output. The input signal $v(t)$ and output signal $u(t)$ are linked via some functional relation

$$u(t) = {}_aF_t[v(\cdot)].$$

Here, F is a response functional and a denotes an initial moment of input signal appearance: $v(t) = 0$ for $t < a$. The lower limit a of the functional can be taken as $-\infty$, which allows to include in consideration signals appeared in remote past or composed of several segments separated by silence intervals. For short, we will write F_t instead of ${}_{-\infty}F_t$.

In the frame of LRT, the response functional F should satisfy the following two requirements.

- It is causal: if $v(t) = 0$ for all $t \le t_1$ then $F_t[v(\cdot)] = 0$ for all $t \le t_1$.
- It is linear: for any pair of functions $v_1(t)$ and $v_2(t)$ of some family of functions $\{v(t)\}$ and any pair of constants c_1 and c_2, the following equality takes place: $F_t[c_1v_1(\cdot) + c_2v_2(\cdot)] = c_1F_t[v_1(\cdot)] + c_2F_t[v_2(\cdot)]$.

This functional can be represented in the following integral form,

$$u(t) = \int_{-\infty}^{t} K(t,t')v(t')dt'. \tag{1}$$

- If the kernel K is invariant with respect to shift, $K(t,t') = G(t-t')$, and $G(t) = 0$ for $t < 0$, we deal with the Standard Linear Response Theory (SRLT).

The response formula (1) takes the form of a convolution integral

$$u(t) = \int_{0}^{t} G(t-t')v(t')dt'. \tag{2}$$

An interesting behavior of discharging current in Panasonic supercapacitors (0.22, 0.47 and 1 F) was revealed in [22,23]. This behavior conforms neither of known models based on equivalent circuit because the response is nonuniform in time. Here, we will show that the response of PMSC-MWCNT is linear and the system is time-invariant.

The invariance of response kernel with respect to shift in time opens a wide field for using such power analytical tools as integral transforms. The Mellin transform casts additional light on the sense of fractional operators. Representing function $G(t)$ via its Mellin transform

$$G(t) = \frac{1}{2\pi i} \int_{c-i\infty}^{c+i\infty} t^{-\sigma}\overline{G}(\sigma)d\sigma,$$

and inserting the result into integral term of Equation (2):

$$\int_{0}^{t} G(t-t')v(t')dt' = \int_{c-i\infty}^{c+\infty} w(\sigma)\, {}_0I_t^{1-\sigma}v(t)d\sigma.$$

Here,

$$w(\sigma) = \frac{\Gamma(1-\sigma)}{2\pi i}\overline{G}(\sigma),$$

and

$$_0I_t^{1-\sigma}u(t) = \frac{1}{\Gamma(1-\sigma)}\int_0^t \frac{v(t')dt'}{(t-t')^\sigma}$$

is an integral operator of complex order $\mu = 1 - \sigma$. Passage from the integral to fractional derivatives can be realized say by means of regularization procedure (derivation of a finite part under $\mu < 0$). After multiplying on the weight factor and integrating over σ, this term yields the fractional derivative of distributed order according to the spectral function $w(s)$ [29,30]. As a result, Equation (2) takes the form

$$_0D_t^{\{w(\cdot)\}}v(t) = u(t).$$

The spectral function $w(\sigma)$ defines the LRT kinetics and links it with dynamics of an open system [28].

The application of fractional equations to the analysis of supercapacitors is usually justified by heterogeneity and complexity of porous electrodes [16,23,31–33]. These equations are often interpreted in terms of anomalous diffusion processes. Anomalous diffusion is usually associated with power law expansion of the diffusion packet, $\Delta(t) \propto t^{\alpha/2}$. The case $0 < \alpha < 1$ corresponds to subdiffusion, and the case $\alpha > 1$ to superdiffusion.

Different specific mechanisms can lead to anomalous diffusion processes [33,34]. The features of anomalous diffusion are often described within the continuous time random walk (CTRW) model. The decoupled version of CTRW model implies that different jump lengths and waiting (trapping) times between two successive jumps are independent random variables. Fractional-order diffusion equation can be obtained as the asymptotic equation of CTRW with power law distributions of random trapping times τ: $P\{\tau > t\} \propto t^{-\nu}$ $(0 < \nu < 1)$. Considering CTRW with space dependent jump probabilities, Barkai et al. [35] derived a time-fractional Fokker–Planck equation (FFPE) for the case, when the mean waiting time diverges. FFPE describes anomalous diffusion in an external force field, it is derived in the following form [35,36],

$$\frac{\partial c(x,t)}{\partial t} = {}_0D_t^{1-\nu}K\frac{\partial}{\partial x}\left[\frac{\partial c(x,t)}{\partial x} + \frac{c(x,t)}{kT}\frac{\partial V}{\partial x}\right], \tag{3}$$

where $c(x,t)$ is particle concentration, $V(x)$ is an external potential, K is a generalized diffusion coefficient, kT is the thermal energy.

Here,

$$_0D_t^{1-\nu}c(x,t) = \frac{1}{\Gamma(\nu)}\frac{\partial}{\partial t}\int_0^t \frac{c(x,\tau)}{(t-\tau)^{1-\nu}}d\tau, 0 < \nu \le 1,$$

is the Riemann–Liouville derivative of fractional order $1 - \nu$ [37]. Fundamental solutions to Equation (3) can be found in [34,35,38].

The fractional diffusion equation for the simplest case $V(x) = 0$ has the following form [39],

$$\frac{\partial c(x,t)}{\partial t} = K\,_0D_t^{1-\nu}\frac{\partial^2 c(x,t)}{\partial x^2}. \tag{4}$$

The Fourier transform of Equation (4) leads to the expression

$$(i\omega)^\nu \tilde{c}(x,\omega) = K\Delta\tilde{c}(x,\omega). \tag{5}$$

The flux density for diffusion described by (4) takes the form

$$J(x,t) = K\,_0D_t^{1-\nu}\frac{\partial c(x,t)}{\partial x}. \tag{6}$$

The subdiffusive generalization of Warburg's impedance for a semi-infinite medium can be written in the form [40]

$$Z(j\omega) = B(i\omega)^{-(1-v/2)}.$$ (7)

Here, B is a frequency-independent parameter.

Based on fractional Equations (3) and (4), different generalizations of diffusion impedances can be obtained for different geometries and boundary conditions (see [32,40–42] and references therein). In our recent paper [43], effects produced by ion subdiffusion in lithium-ion cell are described within the fractional-order generalization of the single-particle model. Here, we use relation of fractional-order impedance to anomalous ion diffusion to interpret the behavior of PMSC.

4. Equivalent Circuit Model

Considering the anomalous diffusion of ions in planar electrodes of PMSC and electrolyte in the interelectrode space, we obtained the following simple equivalent circuit of PMSC (Figure 3). It contains CPEs corresponding to ion subdiffusion in the MWCNT array and, in the interelectrode space, a capacitor C_{DL} and a resistor R_{DL} related to the electric double layer capacity and electrolyte resistance. The corresponding impedance is

$$Z(s) = R + \frac{1}{C_\beta s^\beta} + \frac{1}{C_\alpha s^\alpha + [R_{DL} + 1/(C_{DL}s)]^{-1}}$$ (8)

Fitting measured impedance spectra, we determined parameters of equivalent circuit for samples D33 and A53 at room temperature (Table 1). To fit the data, we used the program "EIS Spectrum Analyzer" [44]. An example of such fitting for D33 sample is shown in Figure 4. The value of parameter β can be explained by the weak influence of ion diffusion in the interelectrode space on the impedance spectrum; therefore, the interelectrode region can be described using a series-connected resistor and an almost "ideal" capacitance. Parameter α is related to ion subdiffusion in MWCNT array. According to Equation (7), value $\alpha = 0.7$ corresponds to subdiffusion parameter $v = 2(1 - \alpha) = 0.6$.

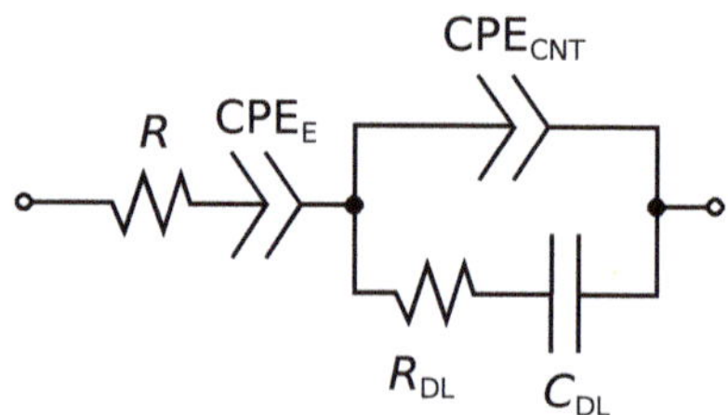

Figure 3. Simple equivalent circuit model of PMSC.

Table 1. Parameters of the equivalent circuit model.

Parameter	D33	A54
R, Ohm	1450	1275
C_{DL}, F	1.86×10^{-5}	1.25×10^{-5}
R_{DL}, Ohm	0	0
C_β, s$^\beta$/Ohm	2.85×10^{-4}	1.01×10^{-4}
β	1.0	0.91
C_α, s$^\alpha$/Ohm	1.14×10^{-4}	1.14×10^{-4}
α	0.70	0.78

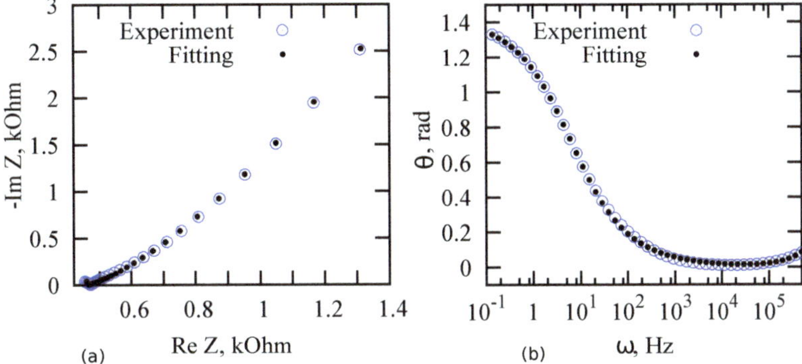

Figure 4. Nyquist plot (**a**) and frequency dependence of phase shift (**b**) for PMSC (D33) with PVA + LiCl electrolyte. Circles are experimental data, dots are the result of fitting by impedance Equation (8).

To find electric current as a response of PMSC, we use the numerical inverse Laplace transformation of the following relation: $\tilde{I}(s) = \tilde{V}(s)[Z(s)]^{-1}$. Note that usually this relation leads to disagreement of numerical results with measured data, because $Z(s)$ depends on operating voltage and it can not be considered as a constant characteristic of the device for all voltages. Moreover, nonlinear effects can take place, whereas $Z(s)$ is obtained for the linearized process.

We use the following input voltage signals and their Laplace transforms.

For potentiostatic charging $V(t)$ and its Laplace transform are

$$V(t) = V_0 H(t), \quad \tilde{V}(s) = \frac{V_0}{s},$$

where H is the Heaviside function. For the discharging process, we have

$$V(t) = V_0[H(t) - H(t - \theta)], \quad \tilde{V}(s) = V_0 \frac{1 - e^{-\theta s}}{s},$$

where θ is a charging time. To plot discharging curves in log-log scale, we shift obtained $I(t)$ in time to charging duration θ.

For cyclic voltammetry, the following voltage signal is used

$$V(t) = 2V_0\left(\frac{t}{\theta} + 2\sum_{k=1}^{\infty}(-1)^k\left(\frac{t}{\theta} - k\right)H\left(\frac{t}{\theta} - k\right) - \frac{1}{2}\right).$$

Its Laplace transform has the form

$$\tilde{V}(s) = 2V_0\left(\frac{1}{\theta s^2} + \frac{2}{\theta s^2}\sum_{k=1}^{\infty}(-1)^k e^{-k\theta s} - \frac{1}{2}\right) = V_0\left(\frac{2}{\theta s^2}\frac{1 - e^{-\theta s}}{1 + e^{-\theta s}} - 1\right) = V_0\left(\frac{2}{\theta s^2}\text{cth}\left(\frac{\theta s}{2}\right) - 1\right).$$

We use the Numerical Laplace Inversion package (Wolfram Mathematica) [45] to invert the Laplace transform $\tilde{I}(s) = \tilde{V}(s)[Z(s)]^{-1}$. Calculated charging and discharging curves, and voltammograms are shown in Figures 5 and 6 and compared with experimental data. There were no fitting parameters, all parameters have been determined from impedance spectra. Thus, we observe quite satisfactory agreement of model responses with experimental data. This means that the PMSC response in potentiostatic and cyclic voltammetry measurements are linear and uniform in time and parameters of equivalent circuit model do not depend on operating voltages.

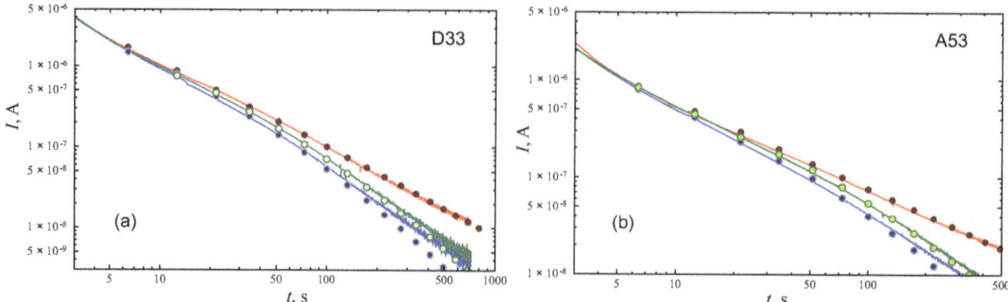

Figure 5. Comparison of measured charging and discharging ($\theta = 100$ and 200 s) curves (lines) for PMSC D33 (**a**) and A53 (**b**) with linear responses predicted by the equivalent circuit model with parameters indicated in Table 1 (points).

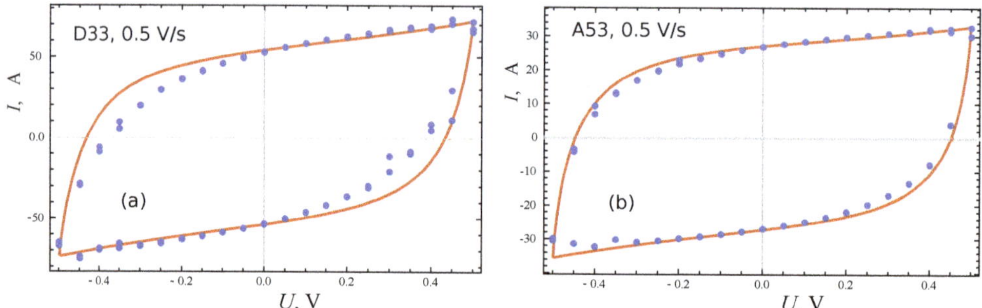

Figure 6. Comparison of measured voltammograms (lines) for PMSC D33 (**a**) and A53 (**b**) with the linear responses predicted by the equivalent circuit model with parameters indicated in Table 1 (points).

Effective capacities of the studied PMSC are estimated to be 102 μF for D33 and 54 μF for A53. D33 is characterized by a smaller electrode stripe width (30 μm) in comparison with A53 (50 μm). The difference in capacities of these samples can be explained by the difference in surface area of contact between the MWCNT array and electrolyte. Despite the fact that the MWCNT array is porous and saturated with electrolyte, it is quite dense and the surface layers of the MWCNT array-electrolyte contact play the main role.

The deviation of the tails of theoretical discharge curves from the experimental data can be explained by variations in the parameter of the constant phase element considered in the next section. These variations can be related to disordering of the array with the height of MWCNT nanotubes.

5. Distributed-Order Fractional Model

To determine the dependence of PMSC capacitance on nanotube height, the MWCNT arrays with similar morphology, but of different heights, were obtained by varying the synthesis time. The dependence of array height on the synthesis time is shown in Figure 7 (left panel). Next, the capacitances of PMSC with PVA+LiCl electrolyte were measured. Figure 7 (right panel) shows the dependence of specific capacity on the height of the array. The capacitive efficiency of using the MWCNT array height h, or ratio C/h, decreases with h. The use of hydrophilic treatments can increase the PMSC capacity by activating the unused area of MWCNT array, however, the power of the device does not increase. Such behavior can be explained by the small variation of the morphology of MWCNT forest for increased heights. These changes can modify the type of ion diffusion inside electrode at different heights (e.g., decrease the subdiffusion exponent). Based on this assumption, we propose distributed-order fractional circuit model and derive corresponding impedances for different distribution of subdiffusion parameter on height.

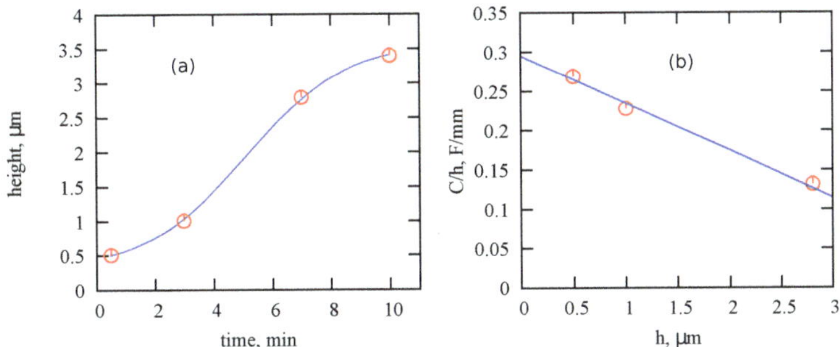

Figure 7. Dependence of MWCNT height on duration of array synthesis (**a**) and the efficiency of using the MWCNT array height for the device capacitance (**b**). Circles correspond to experimental data, lines are approximation by a cubic polynomial function (**a**) and a linear function (**b**).

Consider two cases of distributed-order fractional models for PMSC. The first model implies subdiffusive behavior of fractional order homogeneously distributed over entire volume of electrodes. The second case assumes changing of subdiffusion exponent (fractional order) with height of the array. We show that these models lead to different PMSC impedances and the second case is more preferable to describe our measurements.

5.1. Subdiffusion Exponent Distributed Over Electrodes

For distributed-order fractional model, the subdiffusion equation can be written in the following form [33],

$$\frac{\partial c(x,t)}{\partial t} = K \frac{\partial^2}{\partial x^2} \int_0^1 dv\, \rho(v)\, {}_0D_t^{1-v} c(x,t), \tag{9}$$

where $\rho(v)$ is a probability density function for subdiffusion exponent $0 < v \le 1$, K is a generalized diffusion coefficient.

Equation (9) can be derived from the continuity equation with the following relation between current density and concentration:

$$J(x,t) = -\frac{\partial}{\partial x} \int_0^1 dv\, \rho(v)\, K_v\, {}_0D_t^{1-v}\, c(x,t)$$

Using the Fourier transformation of this distributed-order fractional Fick's law,

$$\tilde{J}(x,\omega) = -\frac{\partial}{\partial x} \int_0^1 dv\, \rho(v)\, K_v\, (i\omega)^{1-v}\tilde{c}(x,\omega),$$

and solution of the diffusion equation in the Fourier domain,

$$\tilde{c}(x,\omega) = C_0 \exp\left[-x\,(i\omega)^{1.2}\left(\int_0^1 dv\, \rho(v)\, K_v\, (i\omega)^{1-v}\right)^{-1.2}\right],$$

we obtain the following impedance,

$$Z(i\omega) \propto \left[i\omega\,\sqrt{\int_0^1 dv\, \rho(v)\, K_v(i\omega)^{-v}}\,\right]^{-1}.$$

For the case of weak dependence of K_ν on ν and uniform distribution of order between ν_{min} and ν_{max}, the impedance takes the form

$$Z(i\omega) = B\sqrt{\frac{\ln(i\omega)}{(i\omega)^{2-\nu_{min}} - (i\omega)^{2-\nu_{max}}}},$$
(10)

where B is a constant parameter.

5.2. Variation of Subdiffusion Order with MWCNT Height

Let us assume the presence of variation in morphology of CNT forest with height of array. Subdiffusion perpendicular to MWCNT inside the electrodes is described by Equation (4), but the order ν depends on height. A separate layer of thickness dh can be characterized by impedance Equation (7). These layers are assumed to operate in parallel, and the equivalent circuit presented in Figure 8 should be used instead of the previous model shown in Figure 3.

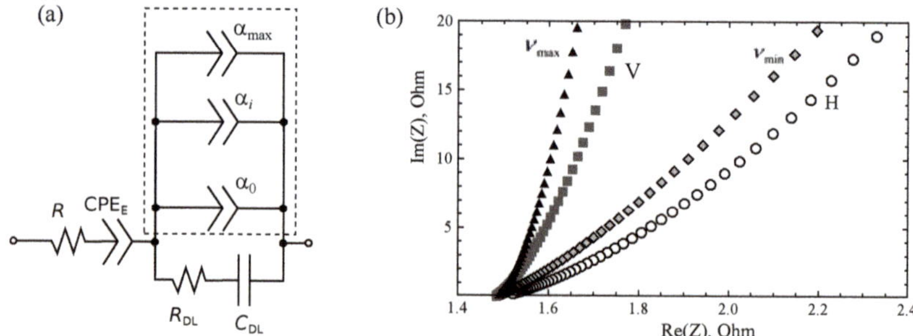

Figure 8. (a) Distributed order fractional circuit model of PMSC. (b) Nyquist diagrams for four cases of fractional PMSC model.

As a result, we obtain the following impedance of anomalous diffusion in electrodes,

$$Z_{dCPE}(i\omega) = \left[\int_{\nu_{min}}^{\nu_{max}} b_\nu(i\omega)^{1-\nu/2}\rho(\nu)\,d\nu\right]^{-1}.$$
(11)

If we assume uniform distribution of ν between ν_{min} and ν_{max} and constant b_ν, we have

$$Z_{dCPE}(i\omega) = \frac{B\,\ln(i\omega)}{(i\omega)^{1-\nu_{min}/2} - (i\omega)^{1-\nu_{max}/2}}.$$
(12)

In Figure 8b, Nyquist plots are presented for four cases of subdiffusion index values. Points denoted as ν_{max} and ν_{min} corresponds to impedance of the circuit presented in Figure 8 with single α-CPE, characterized by $\nu_{min} = 0.5$ or $\nu_{max} = 0.9$. Points denoted by V (volume) and H (height) correspond to the similar circuit including impedances described by Equation (10) or Equation (12), respectively. The shapes of these Nyquist plots look similar, but slope strongly depend on the type of distribution of ν.

Figure 9 demonstrates charging and discharging curves and voltammograms for the circuit presented in Figure 8 with $\nu_{min} = 0.5$, $\nu_{max} = 0.9$. Other parameters are $R = 1275$ Ohm, $C_\beta = 0.101 \times 10^{-3}$ s$^\beta$/Ohm, $\beta = 0.91$, $R_{DL} = 0$, $C_{DL} = 12.5$ μF. Discharging curves indicate the presence of the memory effect, i.e., dependence on the charging prehistory. Cyclic voltammograms of the distributed-order model have the form similar to the single order model. The difference can be seen by variation of ν_{min}. Figure 10 shows two IV-curves for $\nu_{min} = 0.5$ and $\nu_{min} = 0.7$. Parameters are the same as for Figure 9 except v_{min}. Due to effect of disordering, smaller values of v_{min} can be related to higher MWCNT array. For increased height of array, the capacity of PMSC grows, but the dependence of C on h is not direct proportionality. The shape of cyclic voltammogram for the same scanning rate is changed. This shape is closer to rectangle for higher values of v_{min}. This is qualitative explanation of the behavior observed in Figure 7.

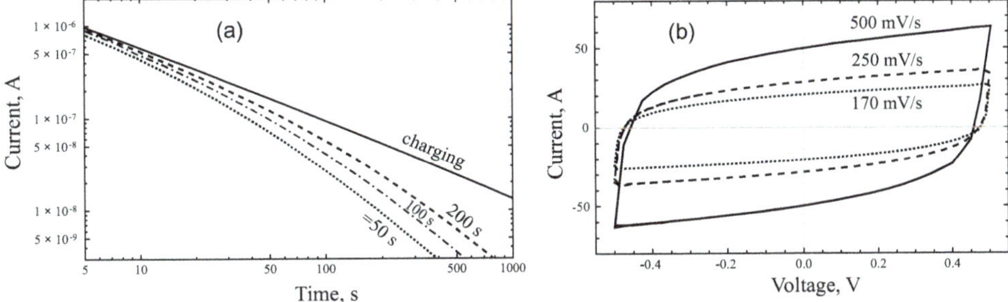

Figure 9. Theoretical charging-discharging curves (**a**) and cyclic voltammograms (**b**) predicted by the distributed-order fractional model.

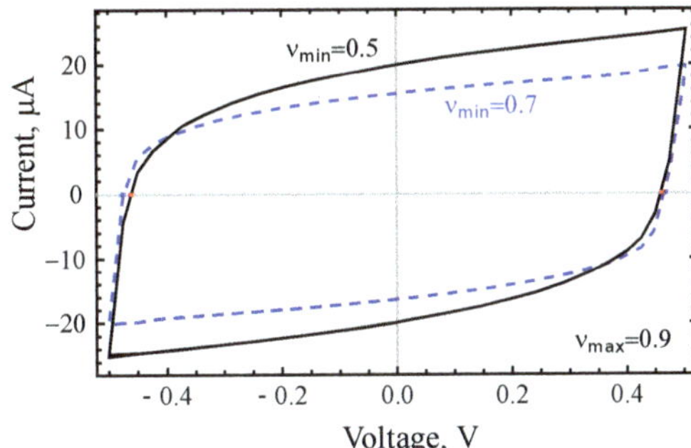

Figure 10. Calculated cyclic voltammograms for the distributed-order fractional model of PMSC for two values of v_{min}.

6. Conclusions

We studied the behavior of PMSC with electrodes of MWCNT array under sinusoidal excitation, step voltage input, and sawlike voltage input. Studied PMSC samples were fabricated on glass substrates by means of standard microelectronics technologies. MWCNTs were synthesized by a catalytic PECVD process. For the first time, a fractional differential model for planar supercapacitor is proposed. The corresponding equivalent circuit model, taking the anomalous diffusion of ions in the array and interelectrode space into account, successfully describes the measured impedance spectra. We demonstrate that the response of the investigated micro-supercapacitors is linear and uniform in time. The numerical inversion of the Laplace transforms for electric current response in an equivalent circuit with a given impedance leads to results consistent with potentiostatic measurements and cyclic voltammograms. Thus, the use of electrodes based on an ordered array of nanotubes reduces the role of nonlinear effects in the behavior of a supercapacitor. In addition, the effect of the disordering of nanotubes with increasing array length on the impedance of a supercapacitor is discussed in the framework of a distributed-order subdiffusion model. The dependence of PMSC capacitance on the height of MWCNT arrays in electrodes is measured. The capacity grows with h, while the ratio C/h decreases. The explanation of such behavior is given within the fractional differential model assuming modification of the array morphology for the increased height that leads to a variation of the subdiffusion exponent.

Author Contributions: Conceptualization, E.P.K., V.V.S. and R.T.S.; methodology, R.T.S.; software, R.T.S.; validation, E.P.K. and R.T.S.; formal analysis, R.T.S. and E.P.K.; resources, V.V.S.; data curation, V.V.S.; writing–original draft preparation, R.T.S.; writing–review and editing, R.T.S. and E.P.K.; supervision, V.V.S.; project administration, V.V.S. and E.P.K.; funding acquisition, E.P.K. and R.T.S. All authors have read and agreed to the published version of the manuscript.

Funding: This work is supported by the Ministry of Science and Higher Education of the Russian Federation through the assignment of 2020 year for Scientific-Manufacturing Complex "Technological Centre". Renat Sibatov acknowledges the financial support from the Russian Science Foundation (project 19-71-10063).

Conflicts of Interest: The authors declare no conflict of interest.

References

1.   Wu, Z.S.; Feng, X.; Cheng, H.M. Recent advances in graphene-based planar micro-supercapacitors for on-chip energy storage. *Natl. Sci. Rev.* **2014**, *1*, 277–292. [CrossRef]

2.   Chen, Y.T.; Ma, C.W.; Chang, C.M.; Yang, Y.J. Micromachined Planar Supercapacitor with Interdigital Buckypaper Electrodes. *Micromachines* **2018**, *9*, 242. [CrossRef]

3.   Hu, H.; Pei, Z.; Ye, C. Recent advances in designing and fabrication of planar micro-supercapacitors for on-chip energy storage. *Energy Storage Mater.* **2015**, *1*, 82–102. [CrossRef]

4.   Shao, Y.; Li, J.; Li, Y.; Wang, H.; Zhang, Q.; Kaner, R.B. Flexible quasi-solid-state planar micro-supercapacitor based on cellular graphene films. *Mater. Horizons* **2017**, *4*, 1145–1150. [CrossRef]

5.   Quintana, J.; Ramos, A.; Nuez, I. Identification of the fractional impedance of ultracapacitors. *Fract. Differ. Its Appl.* **2006**, *2*, 432–436. [CrossRef]

6.   Martin, R.; Quintana, J.J.; Ramos, A.; de la Nuez, I. Modeling of electrochemical double layer capacitors by means of fractional impedance. *J. Comput. Nonlinear Dyn.* **2008**, *3*, 021303. [CrossRef]

7.   Dzielinski, A.; Sierociuk, D. Ultracapacitor modelling and control using discrete fractional order state-space model. *Acta Montan. Slovaca* **2008**, *13*, 136–145.

8.   Bertrand, N.; Sabatier, J.; Briat, O.; Vinassa, J.M. Embedded fractional nonlinear supercapacitor model and its parametric estimation method. *IEEE Trans. Ind. Electron.* **2010**, *57*, 3991–4000.

9.   Elwakil, A.S. Fractional-order circuits and systems: An emerging interdisciplinary research area. *IEEE Circuits Syst. Mag.* **2010**, *10*, 40–50. [CrossRef]

10.   Freeborn, T.J.; Maundy, B.; Elwakil, A.S. Fractional-order models of supercapacitors, batteries and fuel cells: A survey. *Mater. Renew. Sustain. Energy* **2015**, *4*, 9. [CrossRef]

11.   Sibatov, R.; Uchaikin, V.; Ambrozevich, S. Fractional derivative formalism for non-destructive insulation diagnosis by polarization–depolarization current measurements. *J. Vib. Control.* **2016**, *22*, 2109–2119. [CrossRef]

12.   Jesus, I.S.; Tenreiro Machado, J. Application of integer and fractional models in electrochemical systems. *Math. Probl. Eng.* **2012**, *2012*, 248175. [CrossRef]

13.   Westerlund, S.; Ekstam, L. Capacitor theory. *IEEE Trans. Dielectr. Electr. Insul.* **1994**, *1*, 826–839. [CrossRef]

14.   Biswas, K.; Bohannan, G.; Caponetto, R.; Lopes, A.M.; Machado, J.A.T. *Fractional-Order Devices*; Springer: Berlin/Heidelberg, Germany, 2017.

15.   Gil'mutdinov, A.K.; Ushakov, P.A.; El-Khazali, R. *Fractal Elements and their Applications*; Springer: Berlin/Heidelberg, Germany, 2017.

16.   Sabatier, J. Fractional Order Models for Electrochemical Devices. In *Fractional Dynamics*; Sciendo Migration: Boston, MA, USA, 2015; pp. 141–160.

17.   Allagui, A.; Freeborn, T.J.; Elwakil, A.S.; Fouda, M.E.; Maundy, B.J.; Radwan, A.G.; Said, Z.; Abdelkareem, M.A. Review of fractional-order electrical characterization of supercapacitors. *J. Power Sources* **2018**, *400*, 457–467. [CrossRef]

18.   Kopka, R. Changes in Derivative Orders for Fractional Models of Supercapacitors as a Function of Operating Temperature. *IEEE Access* **2019**, *7*, 47674–47681. [CrossRef]

19.   Cahela, D.R.; Tatarchuk, B. Impedance modeling of nickel fiber/carbon fiber composite electrodes for electrochemical capacitors. In Proceedings of the IECON'97 23rd International Conference on Industrial

Electronics, Control, and Instrumentation (Cat. No.97CH36066), New Orleans, LA, USA, 14 November 1997; Volume 3, pp. 1080–1085.

20. Mahon, P.J.; Paul, G.L.; Keshishian, S.M.; Vassallo, A.M. Measurement and modelling of the high-power performance of carbon-based supercapacitors. *J. Power Sources* **2000**, *91*, 68–76. [CrossRef]

21. Wang, Y.; Hartley, T.T.; Lorenzo, C.F.; Adams, J.L.; Carletta, J.E.; Veillette, R.J. Modeling ultracapacitors as fractional-order systems. In *New Trends in Nanotechnology and Fractional Calculus Applications*; Springer: Berlin/Heidelberg, Germany, 2010; pp. 257–262.

22. Uchaikin, V.; Sibatov, R.; Ambrozevich, S. Comment on "Review of characterization methods for supercapacitor modelling". *J. Power Sources* **2016**, *307*, 112–113. [CrossRef]

23. Uchaikin, V.; Ambrozevich, A.; Sibatov, R.; Ambrozevich, S.; Morozova, E. Memory and nonlinear transport effects in charging–discharging of a supercapacitor. *Tech. Phys.* **2016**, *61*, 250–259. [CrossRef]

24. Dzieliński, A.; Sarwas, G.; Sierociuk, D. Time domain validation of ultracapacitor fractional order model. In Proceedings of the 49th IEEE Conference on Decision and Control (CDC), Atlanta, GA, USA, 15–17 December 2010; pp. 3730–3735.

25. Freeborn, T.J.; Maundy, B.; Elwakil, A.S. Measurement of supercapacitor fractional-order model parameters from voltage-excited step response. *IEEE J. Emerg. Sel. Top. Circuits Syst.* **2013**, *3*, 367–376. [CrossRef]

26. Bertrand, N.; Sabatier, J.; Briat, O.; Vinassa, J.M. Fractional non-linear modelling of ultracapacitors. *Commun. Nonlinear Sci. Numer. Simul.* **2010**, *15*, 1327–1337. [CrossRef]

27. Lebedev, E.; Kitsyuk, E.; Gavrilin, I.; Gromov, D.; Gruzdev, N.; Gavrilov, S.; Dronov, A.; Pavlov, A. Fabrication technology of CNT-Nickel Oxide based planar pseudocapacitor for MEMS and NEMS. *J. Physics Conf. Ser.* **2015**, *643*, 012092. [CrossRef]

28. Uchaikin, V. On fractional differential Liouville equation describing open systems dynamics. *Belgogrod State Univ. Sci. Bull. Math. Phys.* **2014**, *37*, 58–67.

29. Chechkin, A.V.; Gorenflo, R.; Sokolov, I.M.; Gonchar, V.Y. Distributed order time fractional diffusion equation. *Fract. Calc. Appl. Anal.* **2003**, *6*, 259–280.

30. Sokolov, I.; Chechkin, A.; Klafter, J. Distributed-order fractional kinetics. *arXiv* **2004**, arXiv:0401146.

31. Allagui, A.; Freeborn, T.J.; Elwakil, A.S.; Maundy, B.J. Reevaluation of performance of electric double-layer capacitors from constant-current charge/discharge and cyclic voltammetry. *Sci. Rep.* **2016**, *6*, 38568. [CrossRef] [PubMed]

32. Sibatov, R.T.; Uchaikin, V.V. Fractional kinetics of charge carriers in supercapacitors. *Appl. Eng. Life Soc. Sci.* **2019**, 87. [CrossRef]

33. Uchaikin, V.V.; Sibatov, R.T. *Fractional Kinetics in Solids: Anomalous Charge Transport in Semiconductors, Dielectrics, and Nanosystems*; World Scientific: Singapore, 2013.

34. Metzler, R.; Klafter, J. The random walk's guide to anomalous diffusion: A fractional dynamics approach. *Phys. Rep.* **2000**, *339*, 1–77. [CrossRef]

35. Barkai, E.; Metzler, R.; Klafter, J. From continuous time random walks to the fractional Fokker–Planck equation. *Phys. Rev. E* **2000**, *61*, 132. [CrossRef]

36. Metzler, R.; Barkai, E.; Klafter, J. Deriving fractional Fokker–Planck equations from a generalised master equation. *EPL (Europhys. Lett.)* **1999**, *46*, 431. [CrossRef]

37. Samko, S.G.; Kilbas, A.A.; Marichev, O.I. *Fractional Integrals and Derivatives*; Gordon and Breach Science Publishers: Yverdon Yverdon-les-Bains, Switzerland, 1993; Volume 1993.

38. Uchaikin, V.V. Self-similar anomalous diffusion and Levy-stable laws. *Physics-Uspekhi* **2003**, *46*, 821. [CrossRef]

39. Balakrishnan, V. Anomalous diffusion in one dimension. *Phys. A Stat. Mech. Its Appl.* **1985**, *132*, 569–580. [CrossRef]

40. Bisquert, J.; Compte, A. Theory of the electrochemical impedance of anomalous diffusion. *J. Electroanal. Chem.* **2001**, *499*, 112–120. [CrossRef]

41. Lenzi, E.; Evangelista, L.; Barbero, G. Fractional diffusion equation and impedance spectroscopy of electrolytic cells. *J. Phys. Chem. B* **2009**, *113*, 11371–11374. [CrossRef] [PubMed]

42. Ciuchi, F.; Mazzulla, A.; Scaramuzza, N.; Lenzi, E.; Evangelista, L. Fractional diffusion equation and the electrical impedance: Experimental evidence in liquid-crystalline cells. *J. Phys. Chem. C* **2012**, *116*, 8773–8777. [CrossRef]

43. Sibatov, R.T.; Svetukhin, V.V.; Kitsyuk, E.P.; Pavlov, A.A. Fractional Differential Generalization of the Single Particle Model of a Lithium-Ion Cell. *Electronics* **2019**, *8*, 650. [CrossRef]

44. Bondarenko, A.; Ragoisha, G. *Progress in Chemometrics Research*; Nova Science Publishers: New York, NY, USA, 2005; pp. 89–102.
45. Valkó, P.P.; Abate, J. Comparison of sequence accelerators forthe Gaver method of numerical Laplace transform inversion. *Comput. Math. Appl.* **2004**, *48*, 629–636. [CrossRef]

Article

Development and Experimental Validation of a Supercapacitor Frequency Domain Model for Industrial Energy Applications Considering Dynamic Behaviour at High Frequencies

Gustavo Navarro *, Jorge Nájera, Jorge Torres, Marcos Blanco, Miguel Santos and Marcos Lafoz

CIEMAT, Government of Spain, 28040 Madrid, Spain; Jorge.Najera@ciemat.es (J.N.); jorgejesus.torres@ciemat.es (J.T.); marcos.blanco@ciemat.es (M.B.); miguel.santos@ciemat.es (M.S.); marcos.lafoz@ciemat.es (M.L.)
* Correspondence: gustavo.navarro@ciemat.es; Tel.: +34-91-335-71-94

Received: 31 December 2019; Accepted: 27 February 2020; Published: 4 March 2020

Abstract: Supercapacitors, one of the most promising energy storage technologies for high power density industrial applications, exchange the energy mostly through power electronic converters, operating under high frequency components due to the commutation. The high frequency produces important effects on the performance of the supercapacitors in relation to their capacitance, inductance and internal resistance (ESR). These parameters are fundamental to evaluate the efficiency of this energy storage system. The aim of the paper is to obtain an accurate model of two supercapacitors connected in series (functional and extrapolated unit) to represent the frequency effects for a wide range of frequencies. The methodology is based on the definition of an appropriate equivalent electric circuit with voltage dependance, obtaining their parameters from experimental tests, carried out by means of electrochemical impedance spectroscopy (EIS) and the use of specific software tools such as EC-Lab® and Statgraphics Centurion®. The paper concludes with a model which reproduces with accuracy both the frequency response of the model BCAP3000 supercapacitors, provided by the manufacturer, and the experimental results obtained by the authors.

Keywords: supercapacitor; energy storage; modelling; frequency model; capacitance; ESR; experimental validation

---

1. Introduction

Among the variety of energy storage systems available for industrial applications, supercapacitors (SCs) have emerged as one of the most selected solutions when high power density, high efficiency, long lifespan, and extended range of operating temperatures are required. Specifically, SCs are widely employed in industry fields where there is a need for storing or releasing a relatively low amount of energy in a very short time and large number of charge-discharge cycles [1,2].

Although SCs are used in many applications by themselves for peak-power pulses reduction or compensation of power oscillations, they can also be hybridized with other energy storage systems, thus increasing the performance of the combined system. Transportation, uninterruptible power supply (UPS), and energy harvesting are the most relevant applications. Firstly, SCs are used in automotive industry to increase the efficiency of hybrid electric vehicles, either for an efficient restart of the engine after a stop, either for limiting the battery current peaks and improving the regenerative breaking in batteries-SCs hybrid vehicles, or for enhancing the dynamic response and load levelling in case of fuel cell-SCs hybrid vehicles [3–7]. Secondly, the UPS industry employs hybrid batteries-SCs systems for reducing the battery stress and extending the lifespan of the system. Moreover, in applications where long-time backups are not required, SCs can totally replace the battery storage system [8–10]. Thirdly, SCs are used in the energy harvesting industry for their integration with non-dispatchable renewable energy sources, i.e., wind and solar energies. The intermittent nature and uncertain prediction of these energy resources results in voltage and frequency fluctuations, which can destabilise the electric grid. Hence, energy storage systems, particularly SCs, play an important role in frequency regulation and peak shaving [11–13].

SCs modelling has become of maximum importance when designing and dimensioning SC installations since it is the way to know in advance about the behaviour and performance of the energy storage system when applied to particular operation profiles or conditions. Control strategies or operational parameters and limits can also be obtained from a model, enlarging the lifetime of the storage technology and therefore achieving a higher level of reliability and competitiveness. Numerous SC models have been published in the literature for different purposes, including capturing electrical dynamic behaviour, which is of utmost importance for the aforementioned industrial applications. The models that capture electrical behaviour of SCs can be classified in three main categories: electrochemical models, intelligent models and equivalent circuit models [2].

Commonly, electrochemical models account for high accuracy and low calculation efficiency, since they capture the reactions inside the SCs by employing coupled partial differential equations (PDEs). Among the electrochemical models, the three-domain model based on the uniform formulation of electrode-electrolyte system developed by Allu et al., and the three-dimensional (3D) model for SCs that considers 3D electrode morphology developed by Wang and Pilon are the most relevant [14,15].

Intelligent modelling techniques, such as artificial neural network (ANN) and fuzzy logic, have the capability of depicting the nonlinear relationship between the performance and its influencing factors, without a detailed understanding of the reactions inside the SCs. Nevertheless, a large amount of training data is needed to ensure model accuracy and generality. The most relevant models in this category are the established ANN model developed by Wu et al., and the one-layer feed-forward ANN developed by Eddahech et al. [16,17].

Equivalent circuit models mimic the electrical behaviour of SCs with parametrised capacitors, inductances, and resistors (RLC). They aim for simplicity, substituting PDEs with ordinary differential equations (ODEs), which hugely facilitate their implementation and make them particularly suitable for industry application analysis and studies. Their accuracy varies depending on the complexity, i.e., number of elements and circuit configuration. In the majority of cases, higher complexity implies higher accuracy.

Three subcategories can be considered for equivalent circuit models: RC models, transmission line models, and dynamic models. The latter, commonly referred to as frequency domain models, account for the best overall performance in terms of complexity, accuracy, and robustness [18]. In this sense, Buller et al. proposed an SC model using electrochemical impedance spectroscopy (EIS) in the frequency domain, with only four experimental parameters [19]. Moreover, Rafik et al. presented a frequency domain model with 14 RLC elements whose values are functions of voltage and/or temperature estimated through EIS methodology [20]. At last, Lajnef et al. developed a less complex frequency domain model for representing the dynamic behaviour of SCs, concluding that SCs present a behaviour which is capacitive at low frequencies, equivalent to a transmission line at intermediate frequencies, and inductive at high frequencies [21].

This paper proposes a new frequency-domain electric model for SCs which considers parameter variations with voltage and self-discharge behaviour. The model improves the overall accuracy of previous frequency-domain models, especially for high frequencies, while keeping a similar degree of complexity and calculation efficiency. A model validation is performed with experimental data collected in the authors' laboratory through EIS tests. The paper is organized as follows: Section 2 presents a review of the most relevant frequency-domain models of SCs. The main contributions of the proposed SC model and its relevance for industry are also highlighted in Section 2. Section 3 describes the experimental tests performed for characterising the SCs behaviour as a function of frequency and voltage, together with the obtained results. The adjustment and description of the SC model is performed in Section 4, together with the comparison between the developed electric model and the collected experimental data. At last, conclusions are discussed in Section 5.

2. Review of the Existing Models of Supercapacitors in the Frequency Domain

When developing a new model that improves one or more features of the existent ones, a comprehensive review of the state of the art is needed. As mentioned in Section 1, the most relevant frequency-domain models reported in the literature are those of Buller et al., Rafik et al., and Lajnef et al. [19–21]. Buller et al. proposed an SC dynamic model using EIS in the frequency domain, which is depicted in Figure 1. The real part of the complex impedance ($\underline{Z}_p$), which represents the porosity of the SC electrodes, increases as the frequency decreases, so the full SC capacitance is only available for DC conditions. Besides, a constant L is included for representing the SC behavior at high frequencies in a simple way, while R_i represents the series internal resistance.

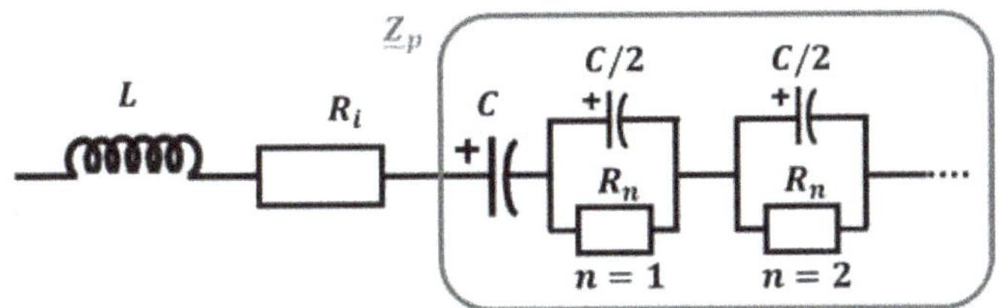

Figure 1. SC frequency domain model proposed by Buller et al (adapted from [19]).

The main contribution of this model is the huge reduction in the number of parameters that need to be determined, diminishing this number to four: L, R_i, C, and the capacitor charge time constant τ (needed to calculate R_n). Moreover, the model takes into account parameter dependency with temperature and voltage, producing precise results. As a counterpart, this dependency requires a large number of EIS measurements, which is time demanding. Moreover, these results are fed into lookup tables, which makes the model non-generalizable and hard to parametrise. Furthermore, the model lacks versatility, since it cannot be used for non-dynamic applications due to the absence of self-discharge modelling. At last, high frequency behaviour is simplified with a constant L, which results in an inaccurate behaviour for frequencies higher than 1 kHz.

The SC model developed by Rafik et al. follows a similar methodology as Buller et al. Parameters are frequency, temperature, and voltage dependent, and are determined from experimental data using EIS. The model is shown in Figure 2.

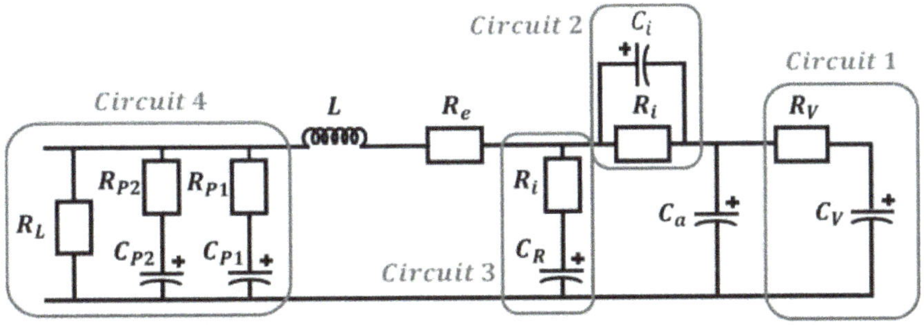

Figure 2. SC frequency domain model proposed by Rafik et al (adapted from [20]).

The SC behaviour is represented by 14 parameters, being L the equivalent inductance for high frequency ranges, R_e the electronic resistance due to all connections, and C_a the principal capacitance. Circuit 1, compounded by R_V and C_V, models the voltage behaviour at low frequencies, while Circuit 2 takes into account the temperature dependency of the electrolyte ionic resistance in the low frequency range. Circuit 3 is introduced in the model to increase the value of the principal capacitance for average frequencies. Finally, Circuit 4 describes the leakage current and the internal charge redistribution, i.e., the self-discharge behaviour of the SC which is also temperature dependent.

The main contribution of the model proposed by Rafik et al. is the detailed behaviour modelling achieved through the inclusion of 3 correction circuits. Furthermore, the model takes into account the leakage current and the charge redistribution on the electrode, making it versatile for a great variety of studies. Moreover, the model is analytical (no lookup tables are required), which makes the model easily generalizable. The main drawback of this model is the huge number of EIS tests that are required to determine the parameters, which is time demanding. Moreover, frequencies higher than 1 kHz, the behaviour is simplified through a constant L value, which outcomes unprecise results.

The frequency domain model developed by Lajnef et al. considers the SC behaviour as a function of frequency and voltage, without considering temperature dependency. Experimental data to model the SC behaviour has also been obtained through EIS tests. The model is displayed in Figure 3.

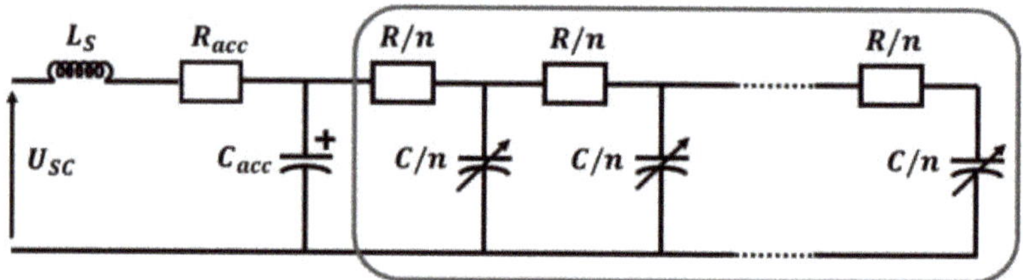

Figure 3. SC frequency domain model proposed by Lajnef et al (adapted from [21]).

The SC is represented by a serial inductor (L_s), an access resistor which physically corresponds to the resistance of the device terminals (R_{acc}), a capacitor (C_{acc}), and a non-linear transmission line with voltage-dependent capacitors. The model considers some simplifications in order to be non-complicated and suitable for frequencies in the range from 10 mHz to 1 kHz: charge redistribution and self-discharge is neglected, high frequency behaviour is not modelled, the impedance real-part dependency on voltage is neglected, and the transmission line is approached by a number of N RC branches. The higher the order N, the better the dynamic response is reproduced by the model.

The main contribution of this model is its simplicity, while keeping a reasonable accuracy for frequencies from 10 mHz to 1 kHz. In addition, the number of EIS tests to perform is low when compared to other models, and the model can be easily parametrized. Furthermore, the model is analytical (no lookup tables are required). As a counterpart, not taking into account temperature and self-discharge effects limits its applicability. At last, SC behaviour for high frequencies is not modelled, making the model non-valid for frequencies higher than 1 kHz.

Considering the drawbacks of the reviewed models, neither of them model the SC behaviour at high frequencies. Although charge/discharge frequency of SCs in industrial applications is usually lower than 1 kHz, high frequency components are always present due to the fact that SCs are integrated with the application by means of power electronic converters. The commutation frequency is in the range of 5 to 50 kHz for IGBT modules, and higher to 100 kHz for MOSFET modules [22]. Besides, industry tends to increase this switching frequency with technologies like silicon carbide. This high-frequency components of voltage and current supplied by the power converter are neglected by the models reported in the literature. Hence, in order to develop a SC model that is suitable for the industry, the high-frequency behaviour of SCs needs to be taken into account.

The model developed in this paper, described in detail in Section 4, takes into account the high-frequency behaviour of a SC, which improves the accuracy for switching frequencies higher than 1 kHz. Moreover, the model is analytical (no lookup tables are required), and has a low number of parameters, which simplifies the parametrisation process and makes it generalizable. The model considers voltage dependence of all parameters, but it does not take into account temperature effects, since, as stated by manufacturers, for temperatures higher than 20 °C, SC properties remain almost constant [23]. This consideration limits its applicability to low temperatures. However, it simplifies the model without compromising the accuracy for operating temperatures from 20 °C to 65 °C. Finally, the developed model includes charge redistribution and self-discharge behaviour, in order to extend its applicability to non-dynamic studies.

3. DC Analysis and Frequency Domain Analysis of Supercapacitors: C and ESR Variations with Frequency

3.1. DC Analysis

The DC characterisation of the supercapacitors (SCs) under experimental analysis is described in this section. The model of SC used to carry out all the experimental tests (DC and frequency domain) is he BCAP3000, commercialised in the past by the Maxwell Company (San Diego, California, USA). The main characteristics of this cell model are summarised in Table 1 [24]:

Table 1. Characteristics of the model BCAP300 cell.

Nominal Capacitance (F)	3000 F
Maximum Capacitance (F)	3600 F
Nominal Voltage (V)	2.7 V
Absolute Maximum Voltage (V)	2.85 V
Maximum Equivalent Series Resistance (ESR_{DC}) initial	0.29 mΩ
Maximum Leakage Current (mA)	5.2 mA
Absolute Maximum Voltage	2.85 V
Short Circuit Current (A)	1900 A
Weight (kg)	0.55 kg

The following paragraph describes the frequency model of the SCs. Regarding the DC analysis, the simplest model of an SC is a capacitor connected in series with a resistance known as equivalent series resistance (ESR). This model is only valid when the charge/discharge current of the SC is continuous, without any AC component. The capacitor represents the element where the energy is stored and the resistance represents the power losses in the SC. Concerning the power losses, it is very important to calibrate them because, the higher the power losses, the lower the efficiency and the higher the temperature reached by the SC. The temperature and the voltage in the SCs are the variables which cause accelerated ageing in this energy storage system. The ageing implies a capacity decrease, a resistance increase and therefore a loss of efficiency. Knowing the resistance in SCs is also important to design a suitable cooling system so that supercapacitors do not reach an excessive temperature.

The capacitance and ESR estimation method used is the one proposed by the manufacturer and it is based on the assumption that EDLC model can be represented by a simple RC equivalent circuit. In [14] it is explained how to calculate both the C and the ESR. Figure 4a shows the voltage profile referred to time applied to SCs to calculate the C and the ESR [15]. As a continuous power source, a Digatron BTS-600 (Digatron, Aachen, Germany), displayed in Figure 4b, has been used to characterise the SCs. This power source allows one to charge/discharge the SC cells from half the nominal voltage until the nominal voltage with a current of 100 A and an error of 0.01%.

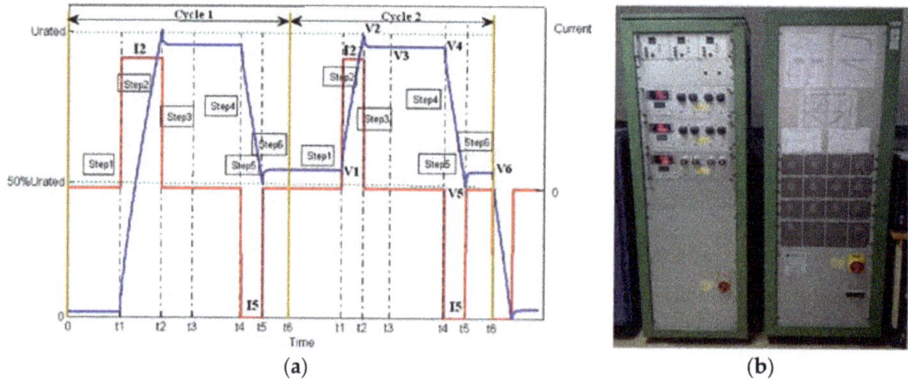

Figure 4. (a) Voltage profile referred to time applied to SCs to calculate the C and the ESR SC; (b) Digatron BTS-600.

SC cells, as well as batteries, operate at low voltage (ranging from 1 to 3 V depending on the chemistry), so that in most cases a series connection of supercapacitor cells is necessary to obtain higher voltage and therefore to provide the required power. For example, a real application operating at a maximum voltage of 650 V would require 240 cells in series with 2.7 V/cell. Thus, it is necessary to take into account in the DC analysis the connection plates. In this regard, the value of this element is calculated before cycling. In this way, a correction must be made to determine the real value of ESR and be able to ignore the electrical connection and measurement contact.

Regarding the capacitance, it should be taken into consideration that SCs are not linear capacitors, but their capacitance depends on the voltage, as shown in Figure 5 [20,25,26]. In the case of cells, manufacturers provide a value of capacity and resistance which is calculated for each cell by using the method mentioned before. Nevertheless, this capacity is just a mean value of capacity.

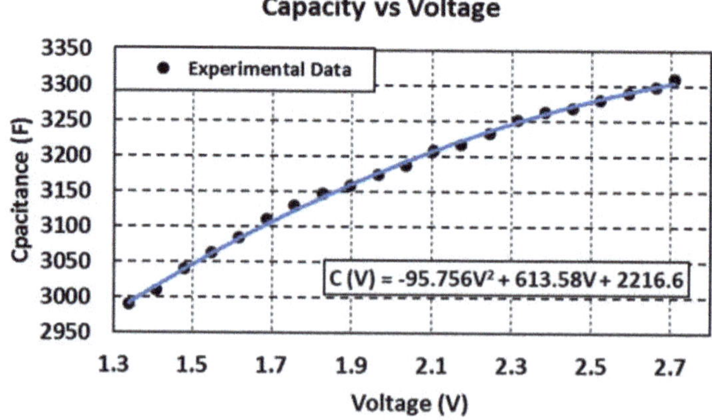

Figure 5. Capacitance evolution as a function of the voltage in the SC.

SCs allow the discharge from nominal voltage down to zero volts, but considering that the stored energy is proportional to the square of cell voltage, they will not be discharged below half of the nominal voltage (1.35 V). Under this condition, they can provide and store 75% of its maximum energy without generating a low voltage ratio with DC link [6].

The nominal capacitance obtained from the datasheet is an average value between the maximum and the minimum voltages of the SC. For this reason, the datasheet gives not only a nominal capacitance value, but a maximum (at the highest voltage) and a minimum (at the lowest voltage) value of this capacitance. At the beginning of the operation life this maximum value is above 3000 F. In order to model this evolution of capacitance, a charging experimental process has been carried out. The average capacitance is calculated with small increases of voltage. Figure 5 shows this evolution. It has a quadratic relationship with voltage as given by the equation below:

$$C\left(V\right) = -95.756 \cdot V^2 + 613.58 \cdot V + 2216.6 \tag{1}$$

Regarding the SC self-discharge effect, the self-discharge current is the current required to keep the SC at a given voltage value. The longer the SC is maintained at a given voltage, the lower the SC leakage current. This effect is influenced by the temperature, the voltage of the cell, the loading history and the ageing. The self-discharge current is given by the manufacturer in the datasheet as a constant value. This value is the charging current required to maintain the SC cell at nominal voltage, after having maintained the SC for 72 h at nominal voltage at a controlled temperature between 23 °C ± 2 °C [27]. This value (2.7 V) is characteristic of the BCAP3000F cell and the maximum value is 5.2 mA, which is calculated after 72 h of assay. As commented before, the leakage current decreases with time, which implies that a conservative estimation is to neglect the variation of the leakage current and set the value to its maximum (5.2 mA in this case).

In order to estimate the leakage current value, an assay has been performed to compare the testing value with the datasheet one. This assay follows the instructions in the technical documentation supplied by the manufacturer [28]. The assay is described in three steps [27]:

- An electric circuit is built, composed by a SC cell, a shunt precision resistance ($R_{shunt} = 1\ m\Omega$) in series with the cell, a jumper in parallel with the resistance, used to charge the SC, and a voltage source. Additionally, an acquisition system is used to record the value of the leakage current over time.
- The cell is charged to its nominal voltage (2.7 V). Then, it is maintained in this state during an hour. After that hour, the jumper is removed and the current crosses through the shunt resistor and is maintained in that state for 72 h.
- After those 72 h, the assay is ended and the last value measured (72 h) is the maximum leakage current. This value is calculated using Equation (2):

$$I_{leakage} = \frac{V_{measured}}{R_{shunt}} \tag{2}$$

Another effect related to the leakage current that occurs during the first 72 h of the assays is the redistribution of the electrical loads. This effect is also relevant for the energy storage system during the whole operation cycle in industrial applications. Assays were carried out on several SC cells (BCAP3000F 2.7 V), and, for the sake of example, experimental results for one of them are shown in Figure 6.

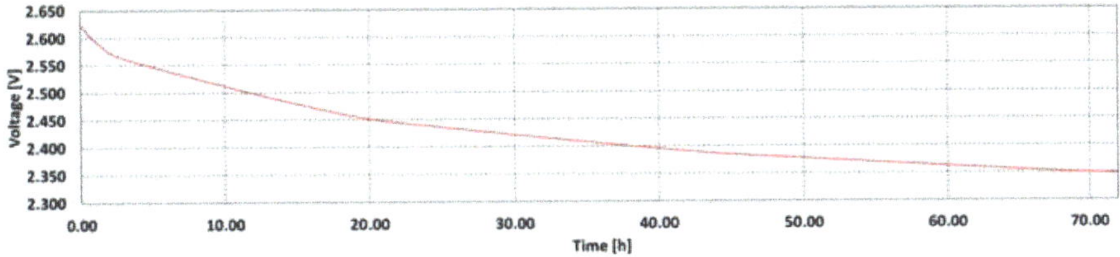

Figure 6. Experimental results of the redistribution of electrical loads and leakage current assay on an SC cell (BCAP3000F 2.7 V) during 72 h after a period of charge.

Using these results, an approximation of the SC voltage evolution with respect to time could be done. As mentioned before, manufacturers provide only the maximum value of the leakage current, which is given to nominal voltage in the supercapacitor. The lower the voltage in the SC, the lower this leakage current. For this reason, the leakage current could be modelled as a variable shunt resistor in function of the voltage in the SC. However, this sole element is not enough to properly represent the leakage current and the process of redistribution of the electrical loads. The electrical circuit that best models the leakage current and the redistribution of the charges is a resistance (R_L) in parallel with a resistance (R_{p1}) connected in series with a capacitor (C_{P1}) as is shown in Figure 7. The capacitance value and the resistance is variable with the voltage. This element connected in parallel to the rest of the electrical circuit model the SC behaviour at very low frequency. The value of these parameters and the rest of the electrical circuit will be discussed in Section 4.

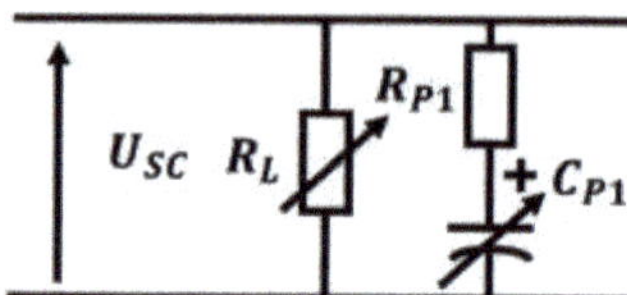

Figure 7. Electrical circuit of redistribution and leakage effects of a SC cell.

In the next section, an electrical circuit model which best fits with the measured frequency response of the SC is proposed. The value of all the elements of the electrical circuit, R_L, R_{P1} and C_{P1} included, for two SCs connected in series will be commented. As will be explained later, the series connection of two SCs will be considered as the minimum functional unit.

3.2. AC Analysis

The details of the procedure to analyse the frequency performance of the SCs are explained below. The experimental data are compared with the data provided by the manufacturer to extract conclusions about the validity of the tests performed and the proposed electrical circuit model. All the experiment results have been obtained for a single cell previously charged to 2.7 V. The state of charge of the supercapacitor is very important because of the variations with the voltage in the SC. The higher the voltage, the higher the capacitance, as shown in the Figure 5. For frequencies below 100 Hz an AC bipolar power source, a BOP 20-20 M s/n E130343 (KEPCO, Flushing, NY, USA) has been used to make the frequency analysis. This power source generates a sinusoidal current signal following the reference from a signal generator. The function generator used is an 33522A s/n MY50005409 (Agilent, Santa Clara, California, USA). A high precision, 34972A s/n MY4901941 datalogger (Keysight, Santa Rosa, California, USA) is also used for the data acquisition to avoid delay between the real and the measured signal. For frequencies from 100 Hz to 100 kHz a HP 4284A Precision LCR meter (Hewlett Packard, San José, California, USA) has been used. In frequencies above 100 Hz, this system provides more precision than the aforementioned power source. According to the switching frequency used in the power electronic converter, which will drive the charge/discharge current through the capacitor, it is not necessary to obtain the response for frequencies above 20 kHz.

In both cases, a shunt resistance with a very low inductance (<10 nH) is used for the measurement of the current. The main electrical characteristics of this resistance are shown in Figure 8. It is a very stable resistance with a 4-wire measurement to avoid measurement errors. The SC is supplied with sinusoidal voltage of 10 mVpeak-peak from 0.001 Hz to 10 kHz with intermediate frequency values (0.001, 0.003, … ., 10 kHz). As in the DC analysis, the connection plates between supercapacitors should be taken into account. A specific test has been carried out connecting two SCs in series.

Resistance	20 mΩ
Tolerance	0.1 %
Temperature Coefficient	< 300 ppm/°C
Temperature Range	55°–140° C
Maximum Power	10 W
Thermal Resistance at Ambient Temperature	< 15° C/W
Thermal Resistance at Aluminium Sustrate	< 3° C/W
Isolation Voltage	500 VAC/VDC
Inductance	<10 nH
Maximum Permanent Current	81 A

Figure 8. Shunt resistance used in the assays; shunt resistance datasheet.

The experimental results obtained for one single cell are shown in red in Figure 9. This result is compared with the experimental data provided by the manufacturer (blue line). Manufacturer data are given for a single BCAP3000 cell charged to 2.7 V at 25 °C. The resistance of the connection plates is not included, so it is necessary to add its value to the provided input data. On the other hand, it is worth mentioning that the manufactured only provides data from 0.013 Hz to 10 kHz, so below and above these frequencies there is no reference values to compare the experimental results against. Finally, the manufacturer only supplies the impedance magnitude versus frequency and the imaginary impedance versus frequency. From these two graphs, the resistance and the capacitance versus frequency evolutions need to be obtained. For frequencies below 0.1 Hz, it is very difficult to calculate the resistance from Maxwell curves without errors because the capacitance (hundreds of Farads) is much higher than the resistance (hundreds of $\mu\Omega$). The experimental resistance and capacitance values are compared in Figure 9a,b respectively. The capacitance value is very similar in both cases: the value obtained experimentally and the value supplied by the manufacturer. However, in case of the resistance, the differences are greater. Especially for frequencies below 0.03 Hz and above 500 Hz. The capacitance value is maximum below 0.01 Hz and it begins to considerably decrease from 0.03 Hz. The resonance frequency is located around 80–100 Hz, where the SC begins to have inductive character.

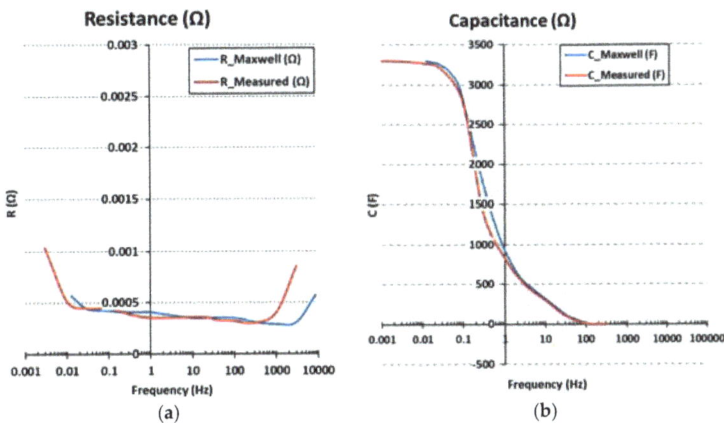

Figure 9. (**a**) SC resistance evolution as a function of the frequency; (**b**) SC capacitance evolution as a function of the frequency. Experimental results and shown in red, while the reference values supplied by the manufacturer are displayed in blue.

As discussed earlier, in industrial applications it is necessary to connect cells of supercapacitors in series to reach a suitable voltage. However, the unequal voltage distribution in the cells will affect the performance and reliability of the stack. This unbalance is due to manufacturing deviations in capacitance, ESR and self-discharge from cell to cell. Self-discharge is a leakage effect dominated by redox reactions at the electrode surface through which electrons cross the double layer [29]. Capacitance variations have a greater effect on the voltage distribution during cycling, and leakage current variations during standby, as in backup applications. Since the voltage distribution between cells is unequal, there may be cells with higher voltage, very close to the upper limit voltage. Furthermore, a high voltage applied to the supercapacitor accelerates the decomposition of the electrolyte, provoking the decrease of capacitance and, as a result, a greater voltage unbalance. Subsequently, an active cell voltage balance or a maximum voltage protection system is required to achieve a good voltage distribution and to prevent any cell from reaching its absolute maximum voltage. The manufacturer supplies an active system to evenly distribute the total voltage between cells. It is not really an active balancing system but an overvoltage protection system. It consists of an electronic board which consumes power when the voltage in the cells is above 2.7 V. Figure 10 shows the connection of SCs in series with the active balance circuit.

Figure 10. Four SCs connected in series with the active balance board in "standby mode" (left one) and in "on mode" (right one).

When the voltage in the cell is above 2.6 V a certain quantity of current drains through a resistor in a voltage balance circuit, presented in Figure 11 connected in parallel with the SC., in order to discharge the cell. The higher above 2.6 V the voltage, the higher this current. The evolution of this current with voltage is shown in Figure 11. The voltage balance circuit response was obtained with a laboratory source.

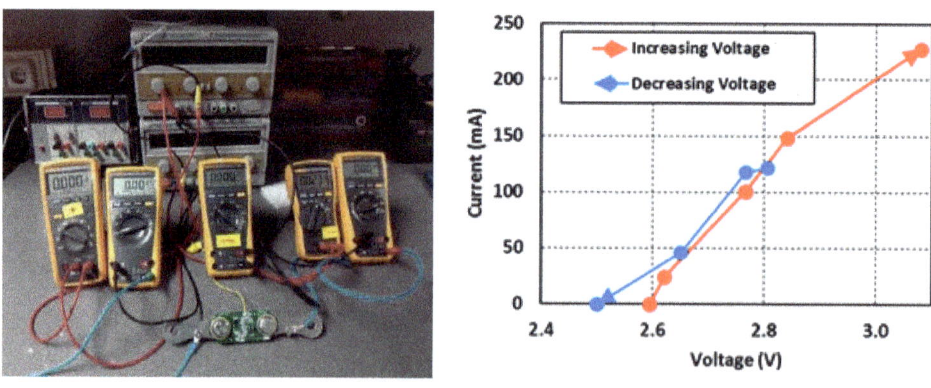

Figure 11. Experimental test to measure the current through the active balance circuit when the voltage in the cell is above 2.6 V.

In order to evaluate the influence of the voltage on the capacitance, the resistance of the connection plates and the voltage balance circuits, four cells of SCs have been connected in series to experimentally determine the frequency response. Four different cases have been studied depending on the preload voltage (V_{bias}) in the SCs (2 V, 6 V, 8 V and nominal voltage 10.8 V). The total measured resistance and capacitance ($R_{TOT} = 4 \cdot R_{SINGLE}$ and $C_{TOT} = C_{SINGLE}/4$) for each case compared with the data supplied by the manufacturer are shown in the Figure 12a,b. The total inductance as a function of the frequency is shown in the Figure 13. As only data of one cell is available from the manufacturer, it is necessary to extrapolate the resistance, the capacitance and the inductance values for four cells connected in series, including the resistance of the connection plates.

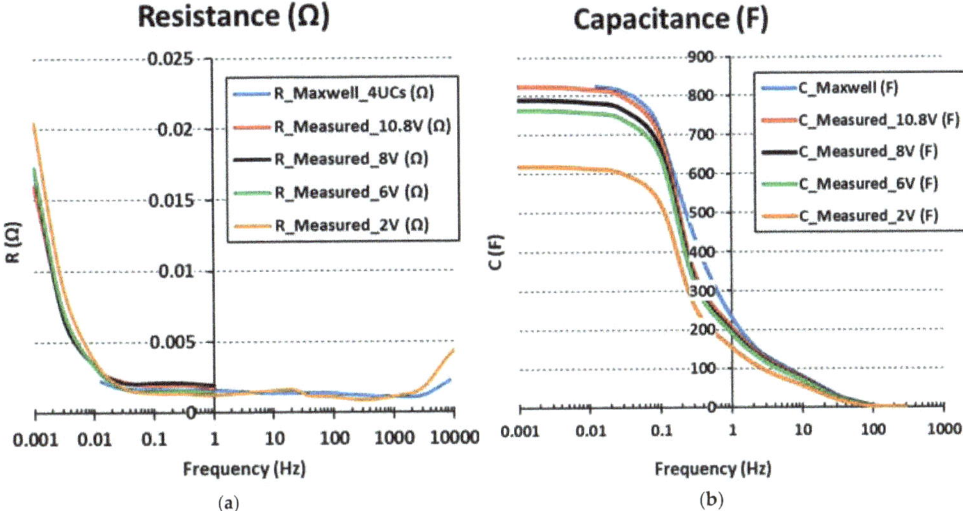

Figure 12. (a) SC resistance and (b) SC evolution as a function of the frequency for different values of initial voltage in the cells.

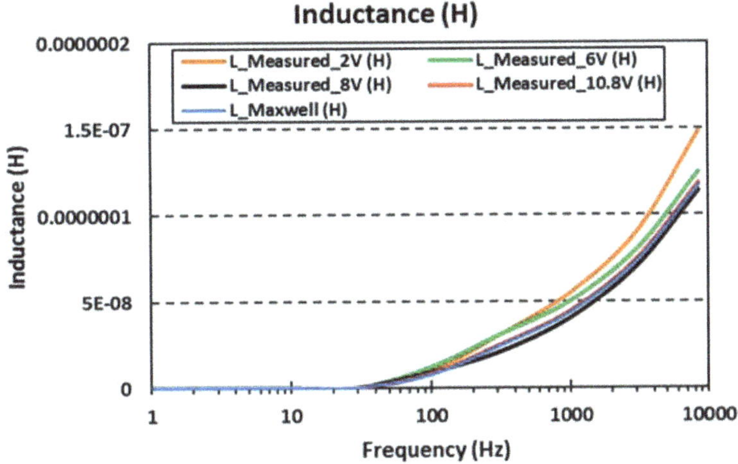

Figure 13. Inductance as a function of the frequency for different initial voltage in the cells.

In this case, the measured resistance is approximately four times greater than the measured value in one single cell. On the other hand, observing the graph one can argue the voltage may have some influence in the resistance in the frequency range below 0.01 Hz. The SCs precharged to 2 and 2.7 V have a very similar resistance value. The resistance of the SCs precharged to 1.5 V is higher than the previous value and the highest resistance value is given in the SCs precharged to 0.5 V. Although it is true that the minimum voltage in operation of the SCs will be 1.35 V, so as not to penalise the efficiency of the power converter which drives the current through the SCs. At half voltage in the SC, only a quarter of the total energy is available. Comparing the capacitance of one single cell with the total capacitance of four SCs connected in series, the former is in the order of four times greater than the latter. Conversely, the total inductance of one single cell is in the order of four times larger than the total inductance value in the series connection of four SCs.

Regarding the value of the capacitance, the higher the voltage in the cells, the higher the capacitance. As already mentioned above, in the voltage range from 1.35 V to 2.7 V the relation between capacitance and voltage is quadratic. Below 1.35 V, the capacitance decreases in a more pronounced way with voltage. As in the case studied of one single cell, the resonance frequency is around 80 Hz.

Figure 13 shows the inductance evolution (sum of the inductances of the four cells) with the frequency. The inductance value varies depending on the voltage in the cells. The higher the voltage, the lower the inductance, although the differences are smaller compared to existing difference in the capacitance. From 80 Hz on, the SC behaves like an inductive element.

4. Frequency Model Design of the Supercapacitor to Adjust the Model Response with the Experimental Response

As explained in the previous section, with electrochemical systems such as SCs, conventional electrical elements (resistances, capacitors and inductances) cannot represent the frequency response of the electrochemical phenomena that occur in this energy storage system. Some of these phenomena are charge and mass transfer and processes of surface diffusion and adsorption. Therefore, equivalent electrical circuits are necessary to model these processes by using in dielectric impedance spectroscopy. These elements are Warburg impedances, constant phase elements (CPE), YAC and ZARC [30–33].

In most of the industrial applications it is necessary the series connection of SCs to reach the suitable voltage levels. For that purpose, a functional unit with two SCs connected in series and an active voltage balance board is proposed as the basic unit to model the series connection of *n* SCs. The complete set of chains of *n* SCs will be the sum of *n*/2 functional units [34]. The suggested model has to fit the experimental data, taking into account the dependence of the capacitance on the voltage. In order to find an equivalent electric circuit to model the behaviour of the series connection of the SCs, the software EC-Lab® (Biologic Science Instruments, Seyssinet-Pariset, France) is used [35]. This software enables the user to compare the frequency response of the SCs obtained experimentally and the selected circuit so that the response of both is the same. Figure 14 shows the equivalent electrical circuit of two SCs connected in series with an active voltage balancing board. This electrical circuit model is very similar to Rafik's model, but it has fewer parameters, eleven as opposed to fourteen. Most electrical circuit parameters in this model have a voltage dependence to get the same frequency response than the frequency response of the SCs obtained from the experimental data. This is because the frequency response of the supercapacitor is variable with the voltage. On the other hand, the electrical circuit proposed models the frequency response of the SCs at high frequencies, above 1 kHz. Most existing SC electrical circuits in the literature, including Rafik's model, do not model the frequency response properly at these frequencies. Taking into account that a power converter will drive the current through the SCs there will be a high frequency component of this current will pass through the SCs with no undesirable effects. This high frequency component will increase the power losses of SCs and the temperature on it provoking an accelerated ageing. On the other hand, the revised models in Section 2 do not describe the methodology to calculate the parameters of the electrical circuit model.

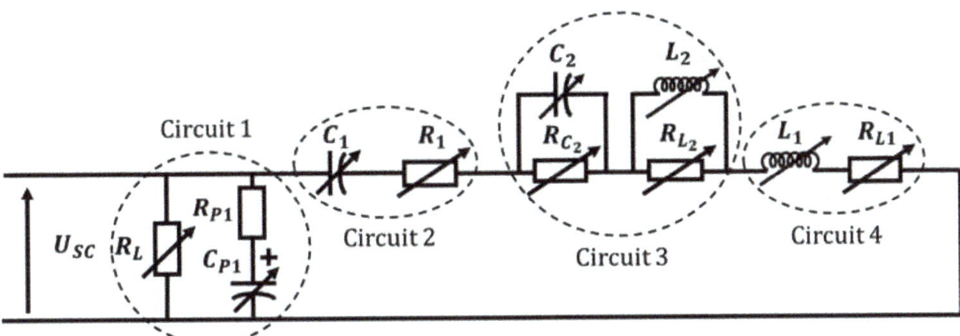

Figure 14. Equivalent electrical circuit to model two SCs connected in series with parameters in function of the voltage in the SC.

As mentioned in Section 3 (DC analysis), R_L, R_{p1}, C_{p1} model the behaviour of the two SCs at very low frequencies, below 1 mHz. The value of these parameters is calculated for very long periods of charge and discharge (very low frequency response). R_1 in series with C_1 represent the behaviour of the SC in the frequency range of 1 mHz to 1 Hz, frequency range where the ESR value provided by the manufacturer is calculated [27]. C_1 is the capacitor with the highest capacitance in the circuit and it is important to set the initial voltage value of this equivalent capacitor in function of the voltage in the two SCs. All parameters in the electrical circuit model are influenced by voltage, but this particular parameter is the one that shows a greater dependence on it. The diffusion phenomena which occurs from 1 Hz to 300 Hz will be modelled with a resistance in parallel with a capacitance ($R_{C2}//C_2$) and a resistance in parallel with an inductance ($R_{L2}//L_2$). In this frequency range, the AC ESR value is found. The first of these circuits ($R_{C2}//C_2$) represents the capacitance zone and the second one represents the inductive zone that is given from the resonance frequency (80–100 Hz). Finally, the resistance R_{L1} connected in series with the inductive L_1 model the behaviour of the SCs at high frequencies.

The value of R_L and C_{p1} is variable with voltage to fit the voltage decay in the SCs with the measured voltage between the positive and negative terminals. As mentioned in Section 3, the total operating voltage of the two SCs is set at between half the nominal voltage and the nominal voltage (2.7 V to 5.4 V). The maximum value of R_L is given at the nominal voltage and the minimum voltage at half the nominal voltage. The value of the capacitor C_{p1} is variable with the total voltage in the SCs (Vbias). R_{p1} is an invariable resistance with voltage of constant value. Table 2 resumes the value of these parameters with voltage. Now, it is necessary to calculate the rest of the electrical circuit elements in function of the frequency response of the SCs.

Table 2. Value of the electrical parameters which model the frecuency response of the two SCs at very low frecuencies. V is the measured voltage between the positive and negative terminal.

R_L (Ω)	R_{P1} (mΩ)	C_{P1} (F)
203.7·V	823.18	93.2028·V

In terms of higher frequency analysis, Figure 15 shows the comparison between the equivalent resistance of the proposed model and the experimental resistance obtained with two SCs connected in series and with an initial voltage of 2.7 V. In the same way, Figure 16 shows the comparision between the evolution of the equivalent capacitance of the electrical circuit model with frecuency and the evolution with frecuency of the real capacitance (from experimental tests).

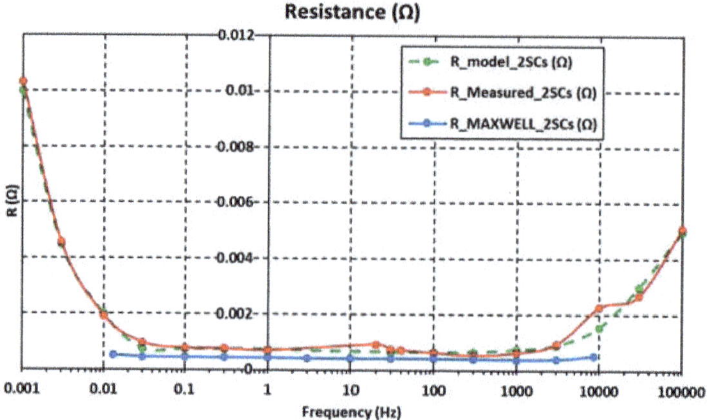

Figure 15. Comparison in the frequency response of the measured resistance from the experimental tests (red line) and the equivalent resistance value from the electrical circuit model of the SCs (green line). The resistance value provided by the manufactured is also plotted in blue.

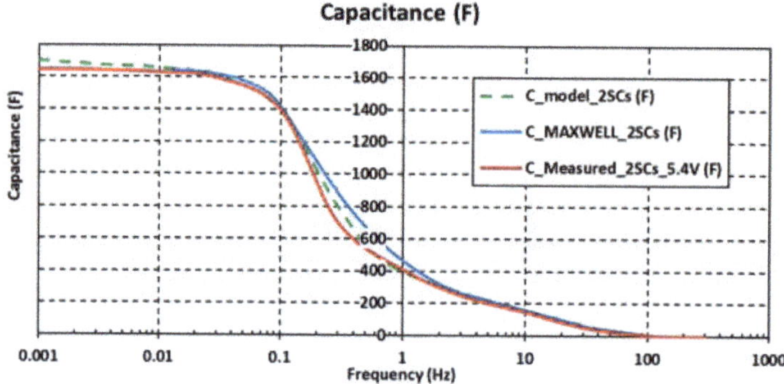

Figure 16. Capacitance value comparison in the frequency response of the electrical circuit (in green) and the obtained value from the experimental tests (in red). Data provided by the manufacturer is also plotted (in blue).

Since the variation of the voltage in the SCs affects the frequency response, it is necessary to calculate the correlation of each circuit element with voltage. In order to derive this mathematical relationship, the value of each element in the circuit is calculated with the software EC-Lab®. Four cases have been studied based on the four experimental tests carried out (2.7 V, 2 V, 1.5 V and 0.5 V in each cell). The circuit parameters obtained for each voltage V_{bias} (sum of the voltage in each SC) are shown in Table 3.

Table 3. Values of the electrical circuit elements for the different initial voltages in the cells.

V_{bias} (V)	R1 (Ω)	C1 (F)	Rc2 (Ω)	C2 (F)	RL2 (Ω)	L2 (H)	L1 (H)	RL1 (Ω)
1	5.45×10^{-4}	708.5	4.94×10^{-5}	1850	4.05×10^{-4}	7.06×10^{-8}	9.29×10^{-8}	6.16×10^{-2}
3	5.74×10^{-4}	1582	6.54×10^{-5}	1152	2.40×10^{-4}	6.04×10^{-8}	8.99×10^{-8}	5.96×10^{-2}
4	5.68×10^{-4}	1644	7.68×10^{-5}	981	2.53×10^{-4}	5.95×10^{-8}	8.25×10^{-8}	5.47×10^{-2}
5.4	5.96×10^{-4}	1801	8.97×10^{-5}	870	2.63×10^{-4}	5.83×10^{-8}	8.02×10^{-8}	5.32×10^{-2}

Subsequently, the software Statgraphics Centurion® is used to determine, by statistical analysis, the relation of all these elements with the voltage (V_{bias}) of the SC. Statgraphics Centurion® is a statistics software that performs and explains the basic and advanced statistical functions [36]. Other softwares can be used to calculate the variation of the electrical parameters with V_{bias} (SPSS®, S-PLUS®, MINITAB®, MATLAB® or simply EXCEL®). Statgraphics Centurion® software is used against other options, but the results obtained with any of the mentioned softwares are very similar. From the analysis carried out, it has been proved that all electrical circuit parameters, to a greater or lesser degree, have a quadratic relationship with voltage. However, the capacitance of the SCs (capacitors of the electrical circuit) has a greater dependence on the voltage than resistance and inductance have. The model equation to calculate any parameter of the electrical circuit in function of the voltage is $A x^2 + B x + C$. The values of these coefficients for each element of the electrical circuit are shown in Table 4.

Table 4. Parameter values of the coefficients A, B y C for the elements of the electrical circuit.

	R1 (Ω)	*C1* (F)	*Rc2* (Ω)	*C2* (F)	*RL2* (Ω)	*L2* (H)	*L1* (H)	*RL1* (Ω)
A	1.12×10^{-5}	−22.96	3.68×10^{-7}	141.02	2.54×10^{-5}	3.10×10^{-9}	-8.98×10^{-11}	1.12×10^{-5}
B	-7.93×10^{-5}	−269.82	1.12×10^{-5}	−1282.12	2.32×10^{-4}	-3.04×10^{-8}	-6.08×10^{-10}	1.12×10^{-5}
C	4.82×10^{-5}	965.2	3.06×10^{-5}	3722.26	5.98×10^{-4}	1.09×10^{-7}	3.93×10^{-8}	1.12×10^{-5}

Figures 17 and 18 show the comparison in the frequency response of the measured capacitance through EIS technique and the electrical model. In both figures, the initial voltage in the cells is different in order to verify that the model fits the real response and it is valid in the entire operating voltage range of SCs.

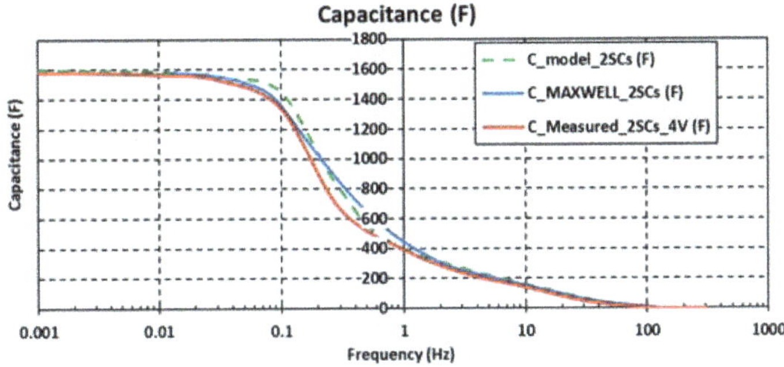

Figure 17. Comparison in the frequency response of the measured capacitance and the value obtained from the electrical circuit model of the two SCs charged to 2 V. The capacitance value provided by the manufacturer is also plotted (blue line).

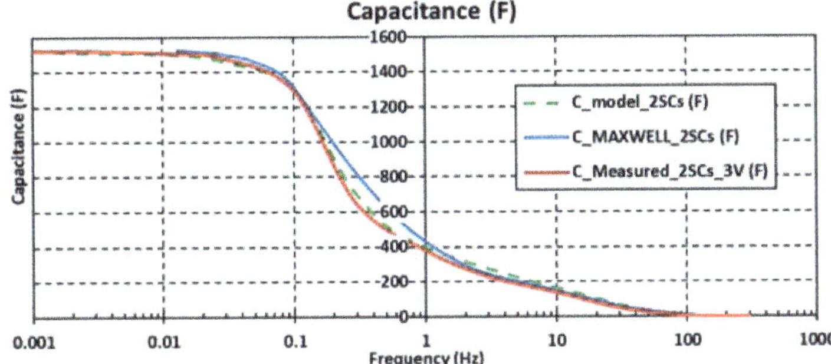

Figure 18. Comparison in the frequency response of the measured capacitance and the electrical circuit model capacitance of the two SCs charged to 1.5 V. The evolution as per the manufacturer is shown in blue.

In the same way as in the case of the capacitance, it is necessary to check if the model fits the real response at high frequencies, where the SC has an inductive behaviour. Figure 19 shows the difference between the SC and the proposed model.

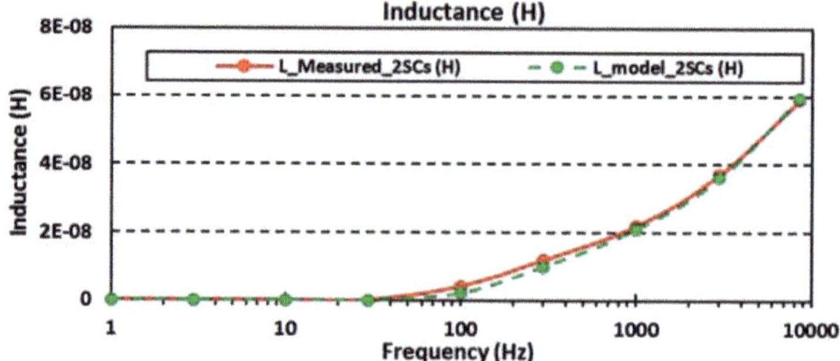

Figure 19. Comparison in the frequency response of the measured inductance and the electrical circuit model of the two SCs charged to 2.7 V.

5. Conclusions

Through the study conducted in this paper, a methodology to determine an electrical circuit which models the frequency response of supercapacitors (SCs) has been described. A basic energy storage system unit, comprising the series connection of two SCs, is proposed. It includes the connection plates and a voltage balancing circuit. The experimental results obtained through EIS methodology have been compared with the proposed model in order to validate it over the entire operation frequency range. This frequency range includes frequencies above 1 kHz, where high-frequency components are found due to the switching frequency of the power electronics. The analysis has evaluated the voltage influence in the impedance of the SCs (ESR, capacitance and inductance). The electrical model can be extrapolated to the series connection of more than two SCs, something very common in the industrial applications. Finally, another conclusion in view of the results obtained is that the faster the charge/discharge of this energy storage system (SC) the lower the capacity value and therefore the lower the energy storage capability.

Author Contributions: G.N. main writing and compilation of paper sections; G.N., J.T. and M.B. developed the models and performed the simulation analysis, G.N., J.T. and J.N. performed the experiments and the analysis of the charging/discharging cycles, J.T., J.N., M.L. and G.N. wrote the paper, M.S. and M.L. complete review of the paper. All authors have read and agreed to the published version of the manuscript.

Funding: This research received no external funding.

Conflicts of Interest: The authors declare no conflict of interest. This research received no external funding.

References

1. Libich, J.; Máca, J.; Vondrák, J.; Čech, O.; Sedlaříková, M. Supercapacitors: Properties and applications. *J. Energy Storage* **2018**, *17*, 224–227. [CrossRef]
2. Zhang, L.; Hu, X.; Wang, Z.; Sun, F.; Dorrell, D.G. A review of supercapacitor modeling, estimation, and applications: A control/management perspective. *Renew. Sustain. Energy Rev.* **2018**, *81*, 1868–1878. [CrossRef]
3. Hu, X.; Jiang, J.; Egardt, B.; Cao, D. Advanced Power-Source Integration in Hybrid Electric Vehicles: Multicriteria Optimization Approach. *IEEE Trans. Ind. Electron.* **2015**, *62*, 7847–7858. [CrossRef]
4. Hu, X.; Johannesson, L.; Murgovski, N.; Egardt, B. Longevity-conscious dimensioning and power management of the hybrid energy storage system in a fuel cell hybrid electric bus. *Appl. Energy* **2015**, *137*, 913–924. [CrossRef]
5. Zuo, W.; Li, R.; Zhou, C.; Li, Y.; Xia, J.; Liu, J. Battery-Supercapacitor Hybrid Devices: Recent Progress and Future Prospects. *Adv. Sci.* **2017**, *4*, 1600539. [CrossRef]

6. Tie, S.F.; Tan, C.W. A review of energy sources and energy management system in electric vehicles. *Renew. Sustain. Energy Rev.* **2013**, *20*, 82–102. [CrossRef]

7. Kouchachvili, L.; Yaïci, W.; Entchev, E. Hybrid battery/supercapacitor energy storage system for the electric vehicles. *J. Power Sources* **2018**, *374*, 237–248. [CrossRef]

8. Aamir, M.; Kalwar, K.A.; Mekhilef, S. Review: Uninterruptible Power Supply (UPS) system. *Renew. Sustain. Energy Rev.* **2016**, *58*, 1395–1410. [CrossRef]

9. Zhan, Y.; Guo, Y.; Zhu, J.; Li, L. Power and energy management of grid/PEMFC/battery/supercapacitor hybrid power sources for UPS applications. *Int. J. Electr. Power Energy Syst.* **2015**, *67*, 598–612. [CrossRef]

10. Lahyani, A.; Venet, P.; Guermazi, A.; Troudi, A. Battery/Supercapacitors Combination in Uninterruptible Power Supply (UPS). *IEEE Trans. Power Electron.* **2013**, *28*, 1509–1522. [CrossRef]

11. Habibzadeh, M.; Hassanalieragh, M.; Ishikawa, A.; Soyata, T.; Sharma, G. Hybrid Solar-Wind Energy Harvesting for Embedded Applications: Supercapacitor-Based System Architectures and Design Tradeoffs. *IEEE Circuits Syst. Mag.* **2017**, *17*, 29–63. [CrossRef]

12. Pegueroles-Queralt, J.; Bianchi, F.D.; Gomis-Bellmunt, O. A power smoothing system based on supercapacitors for renewable distributed generation. *IEEE Trans. Ind. Electron.* **2015**, *62*, 343–350. [CrossRef]

13. Mutoh, N.; Inoue, T. A control method to charge series-connected ultraelectric double-layer capacitors suitable for photovoltaic generation systems combining MPPT control method. *IEEE Trans. Ind. Electron.* **2007**, *54*, 374–383. [CrossRef]

14. Allu, S.; Asokan, B.V.; Shelton, W.A.; Philip, B.; Pannala, S. A generalized multi-dimensional mathematical model for charging and discharging processes in a supercapacitor. *J. Power Sources* **2014**, *256*, 369–382. [CrossRef]

15. Wang, H.; Pilon, L. Mesoscale modeling of electric double layer capacitors with three-dimensional ordered structures. *J. Power Sources* **2013**, *221*, 252–260. [CrossRef]

16. Wu, C.H.; Hung, Y.H.; Hong, C.W. On-line supercapacitor dynamic models for energy conversion and management. *Energy Convers. Manag.* **2012**, *53*, 337–345. [CrossRef]

17. Eddahech, A.; Briat, O.; Ayadi, M.; Vinassa, J.M. Modeling and adaptive control for supercapacitor in automotive applications based on artificial neural networks. *Electr. Power Syst. Res.* **2014**, *106*, 134–141. [CrossRef]

18. Zhang, L.; Wang, Z.; Hu, X.; Sun, F.; Dorrell, D.G. A comparative study of equivalent circuit models of ultracapacitors for electric vehicles. *J. Power Sources* **2015**, *274*, 899–906. [CrossRef]

19. Buller, S.; Karden, E.; Kok, D.; de Doncker, R.W. Modeling the dynamic behavior of supercapacitors using impedance spectroscopy. *IEEE Trans. Ind. Appl.* **2002**, *38*, 1622–1626. [CrossRef]

20. Rafik, F.; Gualous, H.; Gallay, R.; Crausaz, A.; Berthon, A. Frequency, thermal and voltage supercapacitor characterization and modelling. *J. Power Sources* **2007**, *165*, 928–934. [CrossRef]

21. Lajnef, W.; Vinassa, J.M.; Briat, O.; Azzopardi, S.; Woirgard, E. Characterization methods and modelling of ultracapacitors for use as peak power sources. *J. Power Sources* **2007**, *168*, 553–560. [CrossRef]

22. SEMIKRON. *Application Manual Power Semiconductors*, 2nd ed.; SEMIKRON International GmbH: Nuremberg, Germany, 2015.

23. Technologies, M. PRODUCT GUIDE Maxwell Technologies®BOOSTCAP®Ultracapacitors. *Maxwell Technol.* **2009**, *1*, 25.

24. Maxwell. *Product Specifications K2 Series Ultracapacitors*; Maxwell: San Diego, CA, USA, 2014; pp. 1–5.

25. Chaoui, H.; Gualous, H. Online Lifetime Estimation of Supercapacitors. *IEEE Trans. Power Electron.* **2016**, *32*, 7199–7206. [CrossRef]

26. Chaoui, H.; el Mejdoubi, A.; Oukaour, A.; Gualous, H. Online System Identification for Lifetime Diagnostic of Supercapacitors with Guaranteed Stability. *IEEE Trans. Control Syst. Technol.* **2016**, *24*, 2094–2102. [CrossRef]

27. Test Procedures for Capacitance, ESR, Leakage Current and Self-Discharge Characterizations of Ultracapacitors. 2015. Available online: https://www.maxwell.com/images/documents/1007239-EN_test_procedures_technote.pdf (accessed on 2 March 2020).

28. Maxwell Technologies. *Application Note: Test Procedures for Capacitance*; Maxwell: San Diego, CA, USA, 2015.

29. El Brouji, H.; Vinassa, J.M.; Briat, O.; Bertrand, N.; Woirgard, E. Ultracapacitors self discharge modelling using a physical description of porous electrode impedance. In Proceedings of the 2008 IEEE Vehicle Power and Propulsion Conference, Harbin, China, 3–5 September 2008; pp. 8–13.

30. Barsoukov, E.; Macdonald, J.R. *Impendance Spectroscopy: Theory, Experiment and Applications*, 2nd ed.; John Wiley & Son: Hoboken, NJ, USA, 2018.

31. Kanoun, O. *Lecture Notes on Impedance Spectroscopy; Measurement, Modeling and Applications Volume 1*, 1st ed.; CRC Press Taylor & Francis Group: Boca Raton, FL, USA, 2017.

32. Boškoski, P.; Debenjak, A.; Boshkoska, B.M. *Fast Electrochemical Impedance Spectroscopy: As a Statistical Condition Monitoring Tool (SpringerBriefs in Applied Sciences and Technology)*; Springer: Berlin, Germany, 2017.

33. Vadim, A.J.; Lvovich, F. *Applications to Electrochemical and Dielectric Phenomena*; Wiley: Hoboken, NJ, USA, 2012.

34. Dougal, R.A.; Liu, S.; White, R.E. Power and life extension of battery-ultracapacitor hybrids. *IEEE Trans. Components Packag. Technol.* **2002**, *25*, 120–131. [CrossRef]

35. BioLogic Science Instruments. EC-Lab®software-Bio-Logic Science Instruments. Available online: https://www.bio-logic.net/softwares/ec-lab-software/ (accessed on 30 December 2019).

36. Statgraphics. Statgraphics Centurion XVIII. Available online: https://statgraphics.net/ (accessed on 30 December 2019).

Article

Ultracapacitors for Port Crane Applications: Sizing and Techno-Economic Analysis

Mostafa Kermani [1,*]**, Giuseppe Parise** [1]**, Ben Chavdarian** [2] **and Luigi Martirano** [1]

[1] Department of Astronautical, Electrical and Energy Engineering (DIAEE), Sapienza University of Rome, 00184 Rome, Italy; giuseppe.parise@uniroma1.it (G.P.); luigi.martirano@uniroma1.it (L.M.)

[2] P2S, Inc., Long Beach, CA 90815, USA; ben.chavdarian@p2sinc.com

* Correspondence: mostafa.kermani@uniroma1.it

Received: 2 March 2020; Accepted: 7 April 2020; Published: 22 April 2020

Abstract: The use of energy storage with high power density and fast response time at container terminals (CTs) with a power demand of tens of megawatts is one of the most critical factors for peak reduction and economic benefits. Peak shaving can balance the load demand and facilitate the participation of small power units in generation based on renewable energies. Therefore, in this paper, the economic efficiency of peak demand reduction in ship to shore (STS) cranes based on the ultracapacitor (UC) energy storage sizing has been investigated. The results show the UC energy storage significantly reduce the peak demand, increasing the load factor, load leveling, and most importantly, an outstanding reduction in power and energy cost. In fact, the suggested approach is the start point to improve reliability and reduce peak demand energy consumption.

Keywords: ultracapacitor sizing; techno-economic analysis; energy storage system (ESS); ship to shore (STS) crane; peak shaving; energy cost

1. Introduction

It is well known that large-scale commodity (and people) transport uses the sea as a crucial and optimal route. In maritime transportation ports and harbors with power demands of tens of megawatts based on some vast consumers with a high peak level (such as giant cranes, cold ironing, etc.) require a unique power system. Hence, the electrical load areas of port facilities must be organized as microgrids pointing to the goal of a net-zero energy load system that allows a reduced impact on the network supply [1]. The nodes of the maritime highways are the ports representing strategic and critical logistic nodes, both for the strategical service and the energy consumption. Container terminals (CTs) are special zones of the ports used for the transportation of goods as an intermodal exchange between ships and ships and between ships and other transportation vectors (trucks and trains) [2]. Automated terminals are defined as terminals with at least some container handling equipment operating without direct human interaction for 100% of the duty cycle of the equipment. In most cases, drivers have been physically removed from the cranes, although in some cases drivers remain in the equipment cabins but are not needed for the entire duty cycle. There are several types of automated container terminal around the world. The case study of the present paper is Pier E of the Port of Long Beach (POLB), in which the cranes contribute to most of the power demand. The electrical power system in Pier E consists of a subpart of the whole system of the POLB, supplied by an MV/V substation 66/12 kV, owned by Southern California Edison Company [3].

There are different types of crane in operation in a CT. First, ship to shore (STS) cranes, also well known as quay cranes (QC), act as an interface between land and ships and have to load or unload container ships as quickly as possible, usually with multiple cranes working on a single ship. The containers then have to be transported horizontally to the stacking yard behind the STS cranes. This is done by much smaller vehicles like container tractors, automated guided vehicles (AGVs), or automated straddle carriers (ASCs). Finally, the containers are stacked for the most efficient use of space and time, before they are transferred onto trucks or trains that transport them overland. In smaller terminals, this stacking can also be done by straddle carriers, but usually, this is done by rail-mounted gantry (RMG) cranes. Figure 1 provides a flow diagram of a CT operation.

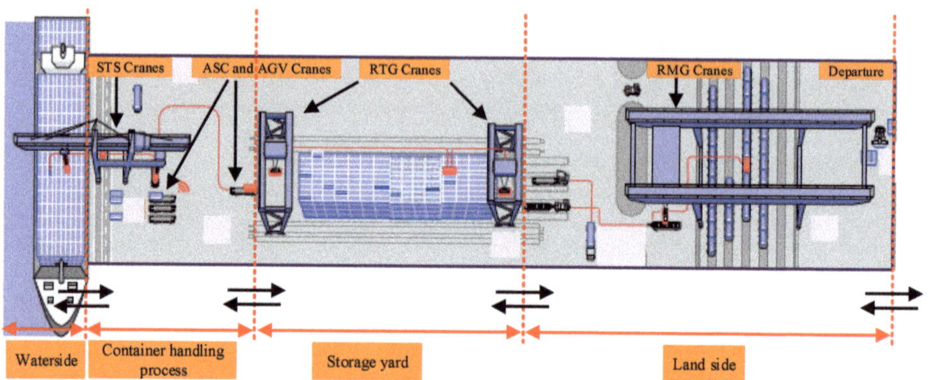

Figure 1. Flow diagram of a container terminal operation.

POLB is one of the busiest and thriving seaports in the world is the second busiest container port in the US, after the Port of Los Angeles (POLA), which adjoins it. With an extension of about 13 km^2, it generates approximately \$100 billion and provides more than 300,000 jobs. In the last few years, as shown in Figure 2, the container trades are reaching the highest level ever, and a contraction has been recorded in a ten-year period [4].

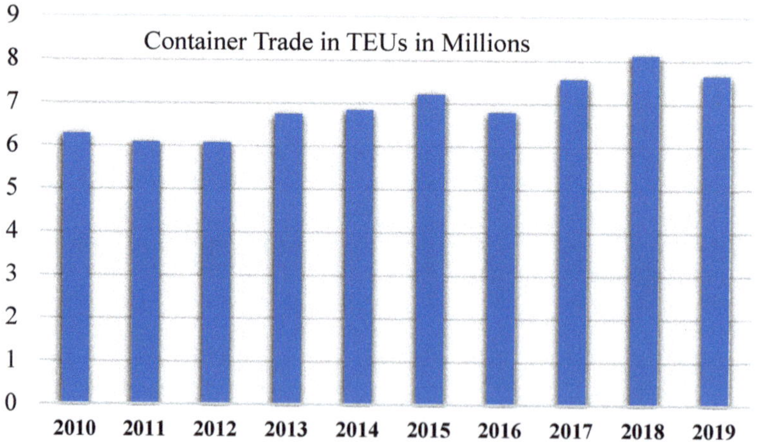

Figure 2. Container trades in TEUs.

Nowadays, by expanding renewable energy sources in electrical power systems, ports authorities have encouraged the use of these units. New solutions for electrical grids are under study in the global community. The main research areas are assigned to renewables, sustainability, low or zero-emissions, smart grids, aggregation, etc. The goal is to create models for smart grids operating in seaports to optimize energy consumption and to reduce the emissions. Not only energy management but also safety and maintenance are the drivers to find new solutions [5,6]. Considering a CT port as a microgrid, such as Figure 3, requires two main infrastructures. The first is the physical assets such as power lines, breakers, transformers, conventional and renewable energy sources, and the second one is communication and control, which is necessary for communication and the controls between the physical assets, for example the local ethernet network, serial connection, microgrid controller, protective relay [7].

Figure 3. Future plan for Energy Island at POLB.

The new scenarios of sustainability, for electrical and energy systems in ports, can be synthesized in the 4-L pillars approach: less, leveled, local, load (also called lele-lolo approach) [8]. Since the cranes, need about 72% of the total power at Pier E (in detail, STS cranes about 37%, ASC cranes 32% and RMG around 3%) [9] with the considerable peak power demand. Hence, the most interesting aspects are the solutions related to the peak-power problem of the cranes. This paper presents a strategy to reduce the peak of the electric power for operation of the group of STS cranes and economic analysis by the integration of energy storage systems (ESSs). The strategy is the optimal sizing of the ESSs, in the first step, and the second one, the coordination of the duty-cycles of the cranes by adopting delay times and synchronization for the operation of the group of STS cranes. The control also exploits the regenerative energy produced by the cranes during some falling phase. The control strategy is based on demand-side management (DSM) with the goal of reduction in the peak (leveled L of 4-L pillars approach). A leveled use of energy aims to have a high load factor and to avoid the billing of peaks unnecessary in the average demand; less use of energy powerfully promotes the rational use of energy and considers the recovery of energy losses as a virtual energy source. Local use of energy promotes a local renewable generation for net-zero energy behavior to reduce environmental pollution and greenhouse gas emissions. Load use of energy promotes an NZELS that avoids reversing energy to the network used as a unidirectional source.

2. STS Crane Operation and Power Demand

In Pier E, double-twin lift cranes are installed with an average consumption of around 3MVA each, during operation with the normal loads, and peak power around 6 MVA for the worst cycle.

Figure 4 shows the one-line diagram for STS cranes power supply at Pier E. Presently, 10 STS cranes do the container moving for two ships at the same time. The development (in the near future) of Pier E is planned for 14 cranes to increase productivity. The minimum number of STS cranes in a group is five, and the total is 10 STS cranes. By this system, a list of different goals as below are achieved:

1. STS cranes' supply by two transformers substation increases reliability dramatically.
2. In the main substation, two independent 66 kV lines with double bus-bar are recommended in which reliability is enhanced.

3.   The double cable lines in the STS cranes' power substations mean that if a fault occurs, the system can reconfigure and operate without delay.

4.   With a double bus-bar, if a failure occurs, it could be isolated, and operation may continue normally on other berths.

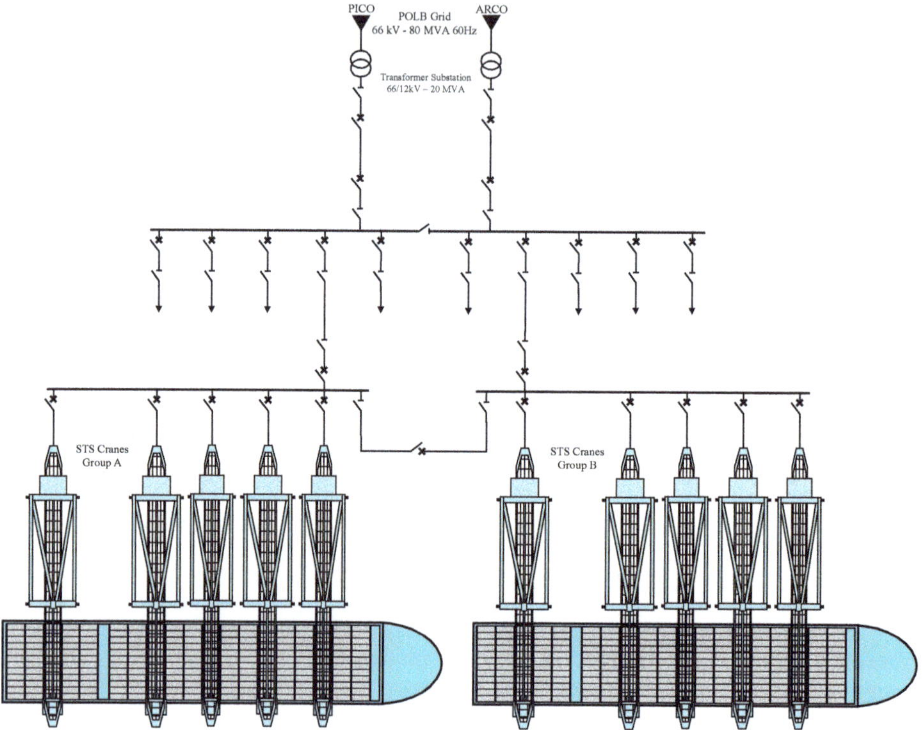

Figure 4. One-line diagram for STS cranes Power Supply.

However, for the mentation of these proposed aims, as a first and main step, it is required to accomplish the cranes' duty cycle operation to reduce the peak power, or regenerative energy flows into the grid from this starting point, thus a flywheel energy storage system (FESS) is proposed [10,11].

An STS crane is able to unload two 40 ft. containers at once, with a cycle that is completed by two independent cranes mounted on the same frame, while sharing a common platform. To differentiate the two cranes, two different acronyms will be used, the first one is ship to the platform (STP, more than one container for lifting) for the main dual hoist crane and the second one for the platform to shore (PTS, only a sigle container for lifting) crane. Since the PTS crane is equipped with a single hoist lifting capability, it has designed to operate two times faster than the PTS sub-crane. Table 1 shows the dimensions, both in feet and meters, of the considered crane, as well as the expected times in which the crane is expected to execute the task of the cycle.

The total time needed for the main crane to complete the operation of the loading and unloading of two containers is around 112 seconds, which depends on a few factors, such as the container's position on the ship, wind, weight, and other aleatory variables. Calculations were made considering a wind speed of less than 5 m/s, standards 65 LT containers, and the main crane in double-lift mode. Figure 5 shows the STS cranes' scheme and speed for a reference duty cycle. The power demand for an STS crane is shown in Figure 6. The green line is for a PTS crane, the blue line for a STP crane, and the red line is the total STS crane power demand. As represented, the power consumption is very high during the main hoist operation, reaching a peak of 3908 kW, and the average power required from the grid is about 378 kW for all duty cycles.

Table 1. The ship to shore (STS) crane's speed and time.

Considered Peration	Meter	Time (s) Load	Time (s) No Load
STP hoist	24.38	20.08	12.38
STP Trolley	64.01	25.84	25.84
STP Lowering	18.29	15.50	10.10
PTS Hoist	9.75	6.63	6.86
PTS Trolley	29.72	12.99	12.99
PTS to AGV	18.29	13.94	10.02

(**a**) A dual STS crane scheme.

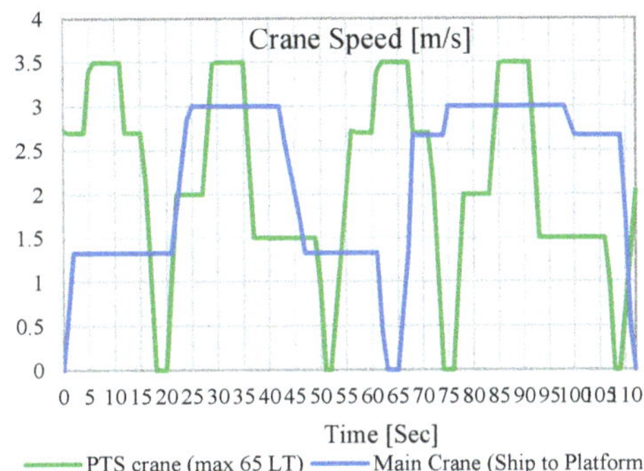

(**b**) Crane speed for a dual hoist STS crane.

Figure 5. STS Crane scheme and reference speed.

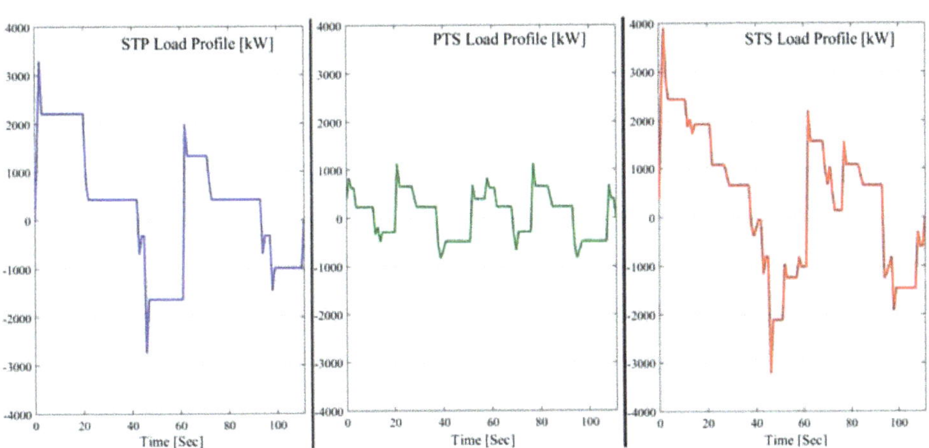

Figure 6. The reference load profile for the different part in an STS crane (Dual Hoist operation).

3. ESSs Selection for Port Cranes

Nowadays, energy storage technologies are used for many applications, for example, power and energy balance, load levelling and peak shaving in power systems. A comparison of the main applications, such as ultracapacitors (UC), battery energy storage systems (BESS) and flywheel energy storage systems (FESS), is shown in Table 2 and Figure 7. UC technologies are proven for different applications, including for peak shaving in cranes. The benefits of power density and fast response time are essential aspects of industrial facilities.

Table 2. Energy storage comparison [12].

Storage	Specific Power (W/kg)	Specific Energy (Wh/kg)	Response Time	Efficiency
UC	500–10000	0.05–15	Milliseconds	>95%
BESS	<500	50–200	Seconds	60%–80%
FESS	400–1500	5–100	Seconds	80%–95%

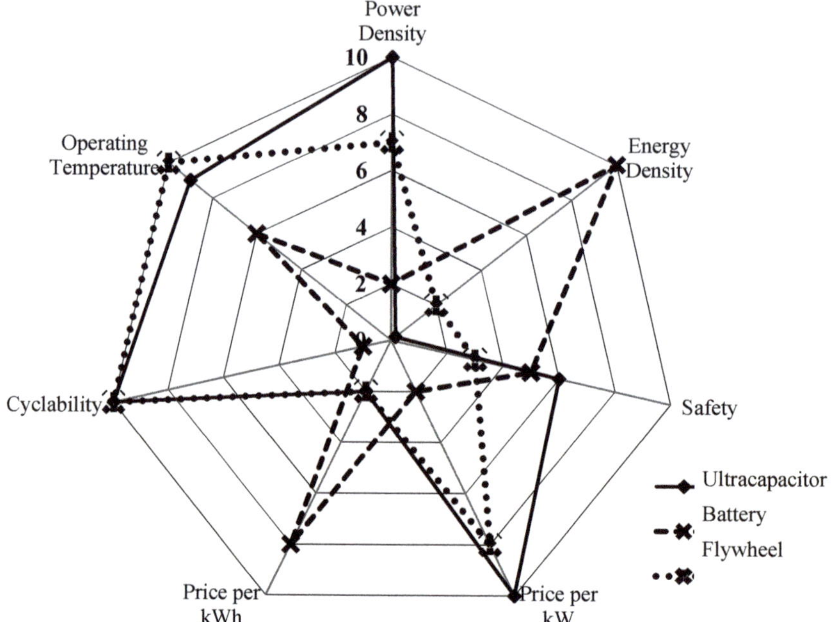

Figure 7. A qualitative comparison between an ultracapacitor, battery, and flywheel.

Due to the fast power changes with the high peak value (in seconds) and independent DC link of each crane, UCs with high power density and fast response time are the best choice for peak power and cost savings. Today, UCs are used in a wide range of applications in industry and transportation [13,14].

Among the most critical applications of UCs can point to controlled electric drives [15,16], RTG cranes and elevators for braking energy [17–21], renewable energy (full cell [22], photovoltaic system [23] and wind turbine [24] contributions), diesel-electric generators include gantry cranes [25–27] and hybrid excavator machines [28], power quality, uninterruptible power supplies, and traction drives (rail, road, and off-road vehicles) [29–32].

4. UC Sizing in Terms of Required Energy and Power

Figure 8 shows the overall scheme of the dual hoist STS crane which is connected to the POLB grid through the AC/DC/AC converters to avoid the voltage rise, leading power factor, and harmonic distortions [33]. The UC bank is connected to the DC link via the bidirectional DC/DC converter. The bidirectional DC/DC converter operates on the boost mode in which electric power is supplied from the UC stage (UC bank discharging in hoisting mode) to the dc link, and on buck mode in which electric power is absorbed from the dc link to the UC (UC bank charging in lowering mode). Bidirectional DC/DC converter applications can handle a high amount of load currents up to several hundred amperes. The control strategy of the DC/DC converter is based on a TMS320vc33 digital signal processor (DSP) which is able to do the speed and electromagnetic torque estimation [34].

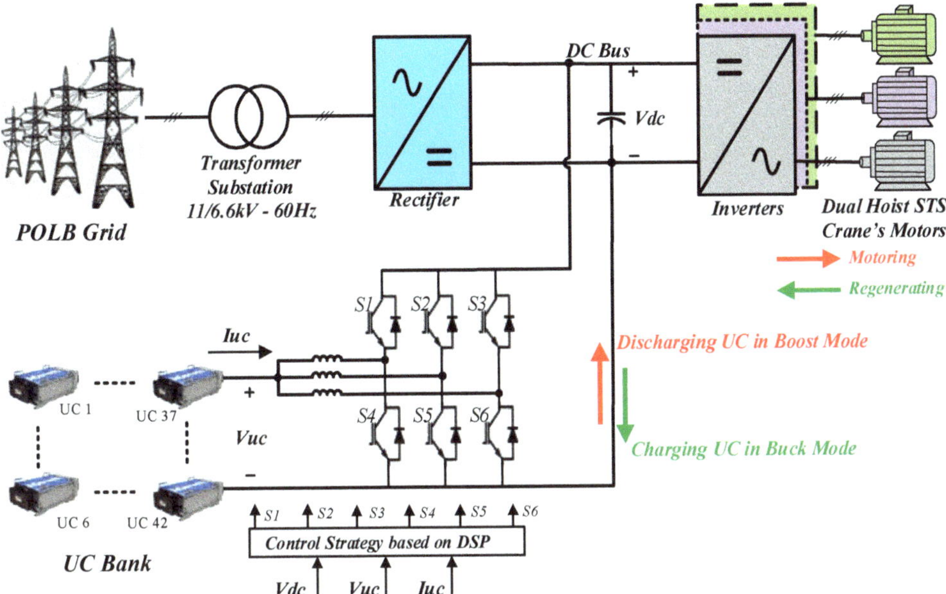

Figure 8. Ultracapacitor (UC) bank location in the dual hoist STS crane via the bidirectional DC/DC converter [35].

The buck-boost converter can be adjusted to keep the output voltage stable because of the varying duty cycle, which can be adjusted by a PWM generator through the internal feedback loop. Based on the required voltage for the motors in the STS cranes (460–470 V), the DC link voltage should be adjusted in such a way by some variations during receiving regenerative power from the load side (UC in charging mode in buck operation) and providing power to the loads (UC in discharging mode in boost operation), the DC link voltage must have upper and lower boundaries revealing the DC link working area.

In this regard, the $V_{DC\,min}$ and V_{DCmax} are obtained by Equations (1) and (2) [36].

$$V_{DCmax} = \frac{2\sqrt{2}}{\sqrt{3}.M_a} \times V_{LLrms} = \frac{2\sqrt{2}}{\sqrt{3} \times 0.95} \times 460 = 800 \text{ V} \tag{1}$$

$$V_{DCmin} = K_r \times V_{LLrms} = 1.35 \times 460 = 621 \text{V} \tag{2}$$

where M_a and K_r are the modulation index and coefficient for a three-phase bridge rectifier, respectively. As the UC voltage plays an outstanding role in power calculation, the upper and lower boundaries of the UC voltage must be declared based on Equations (3) and (4).

$$V_{UCmax} \leq V_{DCmax} \Rightarrow V_{UCmax} = 750 \text{ V} \tag{3}$$

$$V_{UCmin} \geq 0.5 \times V_{UCmax} \Rightarrow V_{UCmin} = 0.5 \times 750 = 375 \cong 400 \text{ V} \tag{4}$$

Figure 9 determines the boundary and limitation for the DC link voltage and UC bank. In this paper, Maxwell BMOD0063 P125 UC module is used for these goals with the rated capacitance 63 F, DC voltage 125 V, rated current 750 A, and maximum power 103.7 kW [37].

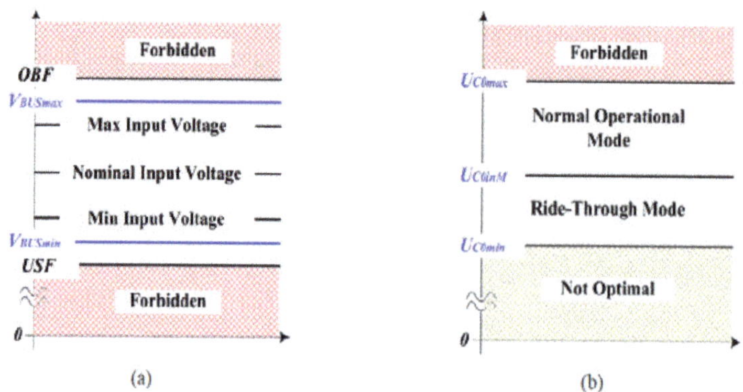

Figure 9. The boundary and limitation for the DC link voltage (**a**) and UC bank (**b**) [38].

As the impact of this approach will increase with the increase in the number of cranes due to the maximum utilization of regenerative energy by other cranes for peak shaving, a strategy should be adopted to minimize both the peak power demand and the excess energy use in the operation of the group with a low number of STS cranes. The peak demand power levels for the dual-hoist STS crane in motoring and generating modes are 3909 kW and −3213 kW, respectively. To achieve the optimal number of UC modules, two different calculations in points of required energy and power is investigated.

4.1. UC Sizing Based on the Energy Production

In the next part, the UC sizing is performed in the required energy and operationg voltage points.

4.1.1. Capacitance Calculation Based on the Required Energy

The total amount of required capacitance is derived from the Equation (5):

$$E_{UC} = 0.5 \times C_{eq} \times \left(V^2_{UCmax} - V^2_{UCmin}\right)$$
$$\Rightarrow C_{eq} = \frac{2 \times E_{UC}}{\left(V^2_{UCmax} - V^2_{UCmin}\right)} = \frac{2 \times 9152 \times 10^3}{750^2 - 400^2} = 45.47 \text{ F} \tag{5}$$

Due to the nominal voltage of the UC module that is 125 V, hence the rated V_{DCmax}, which is adjusted on 750 V, will be achieved by using six UC modules in series connection.

$$N_s = int(\frac{V_{DCmax}}{V_{UC}}) = \frac{750}{125} = 6 \tag{6}$$

By six UCs in series connection, each branch equal capacitance (C_s) carried out in (7).

$$C_s = \frac{63}{6} = 10.5 \text{ F} \tag{7}$$

Consequently, the number of parallel branches which include six UCs (in series) and as a result, the total number of required UC are equal to.

$$N_P = int\left(\frac{C_{eq}}{C_s}\right) = int(\frac{45.47}{10.5}) = 5$$
$$UC_{Total} = 5 \times 6 = 30 \tag{8}$$

4.1.2. Capacitance Calculation based on Voltage Operation Point

According to rate of energy for charging (E_{CH}) and discharging (E_{DCH}) of the UC that are approximately equal to 9152 kW/cycle, the voltage of the UC working point is derived in (9).

$$V_{UCoperation} \cong \sqrt{\frac{E_{DCH}V_{UCmax}^2 + E_{CH}V_{UCmin}^2}{E_{CH} + E_{DCH}}} = \sqrt{\frac{9152 \times 750^2 + 9152 \times 400^2}{9152 + 9152}} \cong 601 \text{ V} \tag{9}$$

According to operation point voltage, the total required capacitance is obtained from (10).

$$C_{eq} = \frac{2 \times E_C}{V_{operation}} = \frac{2 \times 9152 \times 10^3}{601^2} = 50.67 \text{ F} \tag{10}$$

Due to each branch equal capacitance, 10.5 F, the number of parallel branches and the total number of required UCs are calculated by Equation (11) and the UC bank arrangements in terms of required energy is shown in Figure 10.

$$N_P = int\left(\frac{C_{eq}}{C_S}\right) = int\left(\frac{50.67}{10.5}\right) = 5 \tag{11}$$
$$C_{Total} = 5 \times 6 = 30$$

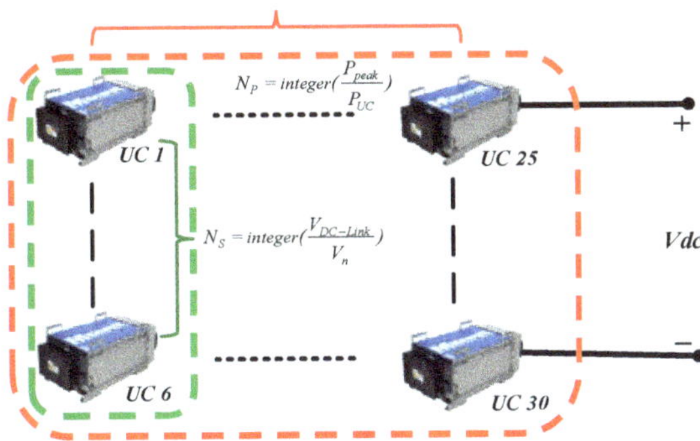

Figure 10. UC bank sizing in terms of required energy.

4.2. UC Sizing Based on the Required Power

As the UC voltage plays an outstanding role in the amount of produced power, the number of required UCs will vary dependently. In this regard, the amount of power produced by each UC module alters between its maximum and minimum boundaries that occur in the case of maximum voltage and minimum voltage, respectively.

4.2.1. P_{UC} when the UC Voltage is Minimum (P_{UCmin})

According to Equation (12) [39], the amount of current that the UC module can produce depends on its state of charge (*SOC*). The equation is as follows:

$$I_{UCmax} = \begin{cases} I_{UCrated} & \text{if } SOC > 50\% \\ (20/3) \times SOC - (400/3) & \text{if } SOC = [20\% - 50\%] \\ 0 & \text{if } SOC < 20\% \end{cases} \tag{12}$$

In this regard, if the $SOC > 50\%$, the generated current can be at the rated value, while for *SOC* of from 50% to 20%, the amount of current proportionately decreases with the *SOC*. Consequently, the *SOC* of all the UCs must remain above 50% in different conditions. Finally, the amount of generated power varies by the alteration of voltage because the current remains steady at the rated amount for $SOC \geq 50\%$. The produced power is derived from (13).

$$V_{UCmin} = \frac{P_{UCmin}}{I_{UCmax}} \Rightarrow P_{UCmin} = V_{UCmin} \times I_{UCrated} \cong 51 \ kW \tag{13}$$

According to the minimum amount of power provided by each UC, that occurs in case of minimum voltage, the total number of UCs is obtained from above:

$$N_{Totalmax} = \frac{P_{Required}}{P_{UCmin}} = \frac{2048}{51} = 40 \tag{14}$$

Due to the placement of six UCs in each branch to provide DC link maximum voltage (750 V), the total number of branches and as a result the total number of required UCs are calculated as follows:

$$N_P = int(\tfrac{40}{6}) = 7$$
$$UC_{Total} = 6 \times 7 = 42 \tag{15}$$

4.2.2. P_{UC} when the UC Voltage is Maximum (P_{max})

The upper limit of generated power is obtained in a case of rated voltage and current as follows:

$$P_{max} = V_{max} \times I_{UCrated} = 103.7 \ kW \tag{16}$$

As a result, the number of required UC is derived from (17).

$$N_{Total,max} = \frac{P_{Required}}{P_{UCmin}} = \frac{2048}{103.7} = 20 \tag{17}$$

Considering six UC modules in each branch, the number of parallel branches and the total required UCs are equal to (18) and the UC bank arrangements in terms of required power is shown in Figure 11.

$$N_P = int(\tfrac{20}{6}) = 4$$
$$N_{Total} = 4 \times 6 = 24 \tag{18}$$

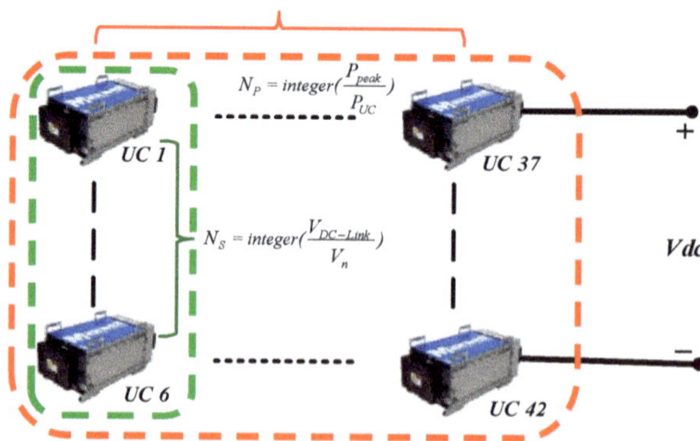

Figure 11. UC bank sizing based on the required power.

In order to increase the reliability of the system, the maximum value of the UC in point on energy and power evaluation should be selected, which carries out 42 UC modules in the Pmin scenario. In the next part, the cost-benefit analysis will be investigated.

5. Cost Benefits Analysis by UC Utilization

Peak shaving based on demand-side management and the integration of energy storage systems is the main step to reduce the cost of electricity, especially for industrial loads. For this goal in the STS crane, this paper considers a hierarchical control strategy to include all the modes. As the first solution, it can be done by coordinate cranes' duty cycles for the STS crane installed in the same meter. This solution is made possible in a full automation CT. Hence, the UC sizing can decrease the local peak for each STS crane and then by duty cycle coordination for the group of STS cranes the general peak reduction has been done.

Cost Benefits for a Single STS Crane

To calculate the first invoice, Southern California Edison, the distribution operator, installs a meter at the beginning of the customer's line, immediately before the point of delivery. The peak power recorded in a month is used to calculate "facilities demand"; the average power consumption is not involved in this calculation. The exposed cost of the energy is around 0.18 $/kWh in the USA, while the exposed cost of peak power demand is about 22 $/kW. According to Table 3, the amount of energy in motoring mode is about 91,512 kW in each cycle, of which any proportion of that can be used to reduce the energy needed to support peak values because the energy's peak usually accounts for around 40% of the total electric bill per month in the US. In terms of financial issues, more peak power reduction causes more money-saving and cost-efficiency. In this regard, in the case of using 1% of the total available energy, the amount of obtained cost is about 20 k$ per month. Also, for 5%, 10%, and 20% energy usage, the amount of earned profit will be about 38 k$, 45 k$, and 53 k$, respectively.

Table 3. Percentage of the peak shaving and monthly earnings based on UC bank for STS crane.

Total ED [kWcycle]	Peak Demand	% ED	The Amount of Peak Shaving	$\%P^{\wedge}S_{UC}$	Monthly Earned Profit ($)
		1	915.12	23.4	20,132
		5	1721.42	44	37,871
91512.93	3908.52	10	2048.82	52.4	45,074
		15	2262.72	57.9	49,779
		20	2441.52	62.5	53,713

Table 4 shows the results for the different percentages of regenerative energy utilization in the generator operation of an STS crane. According to Table 4, the final result for the rate of energy that the UC bank should store is equal to 18.6% of the total regenerating energy, in the case of 10% peak shaving of the total demand power in motoring mode (Table 3). The calculations are as follows:

$$\%P^{\wedge}S_{UC} = \left(\frac{P_{peak-UC}}{P_{Peak-Motoring}}\right) \Rightarrow \%P^{\wedge}S_{UC} = \left(\frac{2048.8}{3908.5}\right) = 52.4\%$$
$$\%P^{\wedge}S_{UC} = \left(\frac{P_{peak-UC}}{P_{Peak-Regeneration}}\right) \Rightarrow \%P^{\wedge}S_{UC} = \left(\frac{1977.8}{3212.6}\right) = 61.5\% \tag{19}$$

Table 4. Percentage of the peak shaving based on UC bank for the STS crane.

Total RE [kWcycle]	Peak Rate	% RE	The Boundary for RE Utilization	$\%P^{\wedge}S_{UC}$
		1.86	915.1	28.4
		9.3	1622.1	51.7
−49,200.3	−3212.6	18.6	1977.8	61.5
		27.9	2175.1	67.7
		37.2	2340.2	72.8

As can be observed in Figure 12, by the optimal UC bank sizing (from a technical and economic point of view), the power demand significantly dropped, as shown in blue. In Figure 12, the red curve is related to the conventional STS crane, the green one is the UC bank charging and discharging operation, and finally, the blue curve is the optimized load profile for the STS crane, and it has been considered as a new load profile for the next step, which will be crane's duty cycle management.

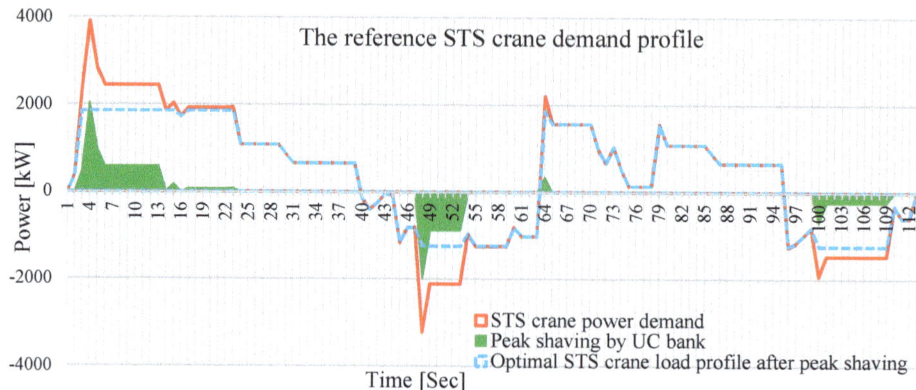

Figure 12. The shaved peak by UC bank to achieve the optimal power profile for the STS crane in kW.

6. Peak Shaving Optimization for the Group of STS Cranes

After UC sizing, to find the best delay times between STS cranes, the particle swarm optimization (PSO) algorithm was applied [40]. The PSO algorithm is a solution for optimization issues such as economic dispatch, control, and losses [41]. It is described as:

1. position (*Xi*) and velocity (*Vi*) vectors are randomly initialized with the problem dimensions. The position of each particle is based on Equation (20) based on the best particle search, the best overall band flight experience, and the velocity vector of its particle.

$$v_i^{k+1} = wv_i^k + c_1r_1\left(Pbest_i^k - x_i^k\right) + c_2r_2\left(Gbest^k - x_i^k\right)$$
$$x_i^{k+1} = x_i^k + Cv_i^{k+1} \tag{20}$$

where c_1 and c_2 are two positive constant integers, r_1 and r_2 are two random numbers with uniform distribution in the range [0~1] and w is the inertia, which is chosen as follows [20,21]:

$$w = w_{max} - \frac{w_{max} - w_{min}}{iter_{max}} \times iter \tag{21}$$

where $iter_{max}$ is the maximum number of repetitions, and $iter$ is the number of repetitions.

2. Fitness of each Pbest particle is measured, and the particle that has the best *Gbest* fit is stored. $Pbest_i^k$ is the best position of the ith particle, which is based on the experience of the particle. *Gbest* is the best position of the particle, based on the group's overall experience. Figure 13 shows the delay time management of the PSO flowchart for the STS crane group.

By the duty cycle management for the group of STS cranes, the normal and optimal load profile with the integration of the UC bank are the reference load profile. It is not common to use only one crane in maritime applications, but port managers used to take advantage of cranes as a group of a few ones. The minimum and the maximum number of STS cranes in the group in Pier E at POLB have been considered, respectively, as five and 10. Consequently, financial calculations are accomplished in a case of using a group of cranes. Table 5 represents the Pmax in different scenarios.

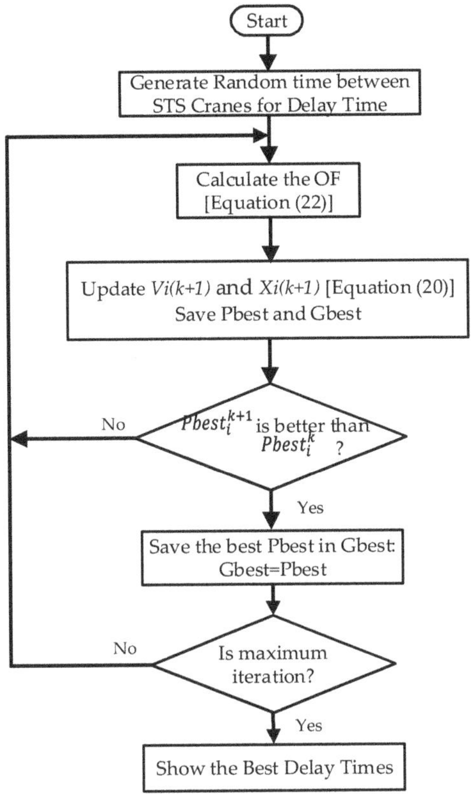

Figure 13. Particle swarm optimization (PSO) flowchart for power demand optimization in the STS crane group.

Table 5. Rated Pmax in different scenarios for the group of STS cranes [kW].

Number of STS Cranes	Duty Cycle Management Scenario		
	Normal Scenario	DSM Scenario	DSM + UC Scenario
5	10,300	5880	4222
10	14,380	8180	5476

Although each crane has a peak power of 4 MW and it is predicted that the system experiences 20 MW as a peak power of a five-crane group, in normal operation the peak power level is about 10 and 14 MW for using five and 10 cranes, respectively. In addition, the system with the only DSM strategy and UC along with DSM strategy scenarios result in almost 5880 kW and 4222 kW of peak power, in a case of using five cranes. Table 6 shows the peak factor rate ($P\hat{}F$), which has been determined with Equation (22):

$$OF(C) = \sum_{i=1}^{n}\left(\frac{P_{max}}{P_{ave}}\right)_i = \sum_{i=1}^{n}\left(P^\wedge F\right)_i \tag{22}$$

where C is a vector of optimization variables and n is the number of cranes, OF is the objective function, P_{max} and P_{ave} are the maximum and the average power of the STS cranes in each iteration in MW, and finally, $P\hat{}F$ is the peak factor rate ($P\hat{}F = Pmax/Pave$).

Table 6. Rated peak factor ($P^\wedge F$) for the STS crane group in different scenarios.

Number of STS Cranes	Duty cycle management Scenario		
	Normal Scenario	DSM Scenario	DSM + UC Scenario
5	5.43	3.1	2.23
10	3.8	2.16	1.45

Table 7 is related to the peak shaving percentage (*%P^S*), which is determined with Equation (23). Peak load shaving is a process that aims to flatten the load curve. Due to the available strategies for load peak shaving, DSM and integration of energy storage technology have been applied for peak reduction in this paper.

$$P^\wedge S_{sc_i} = 1 - \left(P^\wedge F_{sc_i} / P^\wedge F_{Normal}\right) \times 100 \tag{23}$$

Table 7. Percentage of the peak shaving (*%P^S*) for the STS crane group in different scenarios.

Number of STS Cranes	Duty Cycle Management Scenario		
	Normal Scenario	DSM Scenario	DSM + UC Scenario
5	0	43	59
10	0	44	62

Table 8 shows the advantages and disadvantages of using DSM and UC methods from the port authorization point of view. The DSM method has a simple implementation, resulting in cost reduction, while the UC method provides flexible and efficient operation in comparison with the former method

Table 8. Demand-side management (DSM) and UC pros and cons.

	Only DSM Strategy	Both UC and DSM Strategies
Pros	➢ General peak shaving. ➢ Fewer maintenance costs. ➢ Without the required initial cost.	➢ Higher efficiency even in temporary conditions. ➢ Short term payback. ➢ Suitable and efficient operation.
Cons	➢ Not comfortable efficiency in low count condition. ➢ In some forced times, the port authority is not willing to follow. ➢ Complex coordination and operation.	➢ Complicated control strategy. ➢ Required initial investment.

Figure 14 provides an overall scheme of hierarchical STS load reduction which is applied in three different stages. First, local peak shaving is accomplished by using UC for each STS crane. In the second stage, between two STS crane groups, the general peak shaving is carried out by considering a duty cycle recognizing a specific working period for each crane. In the last stage, a flywheel energy storage system (FESS) is used in a common bus between two STS crane groups to fulfill load leveling.

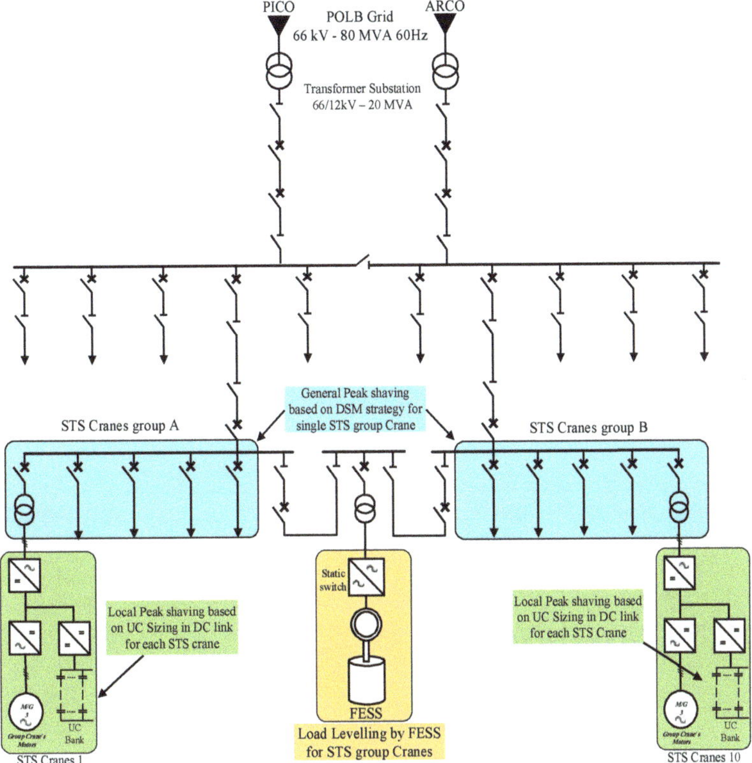

Figure 14. Hierarchical control of for peak load shaving in STS crane group.

7. Cost Evaluation

Financial issues, such as payback duration and total imposed costs, remain one of the most outstanding factors in the initial designing and evaluating the systems. As mentioned before, cost analysis must be carried out in a case of using a few in a group because it is common to select from 5 to 10 cranes as a group in maritime affairs rather than using a single crane. Payback duration will be calculated according to DC/DC cost, which is about 60 k\$, and UC module, cost which is 5 k\$, and due to using 42 UCs for each crane. As five cranes are used, the initial cost is obtained as being 1350 k\$ for each crane.

According to Table 5, the amount of peak shaving which is accomplished by five STS cranes and by using both the UC and DSM scenarios is about 6078 kW, while this amount is almost 4420 kW for using the DSM strategy alone. According to the fact that DSM scenario cost is negligible is and the earned profit must be calculated by a comparison between DSM only and UC along with DSM scenarios. Based on the tariff of imposed cost by peak power, which is 22 \$/kW, the amount of earned profit is about 37 k\$ for a single month by using both the UC and DSM strategies. Due to a one-month maintenance period, the total crane operating duration is considered as 11 months, and the annual earned profit is almost 401 k\$ and the initial investment will be paid back after three years and half. As shown in Figure 15, Pier E will start to profit from the fourth year.

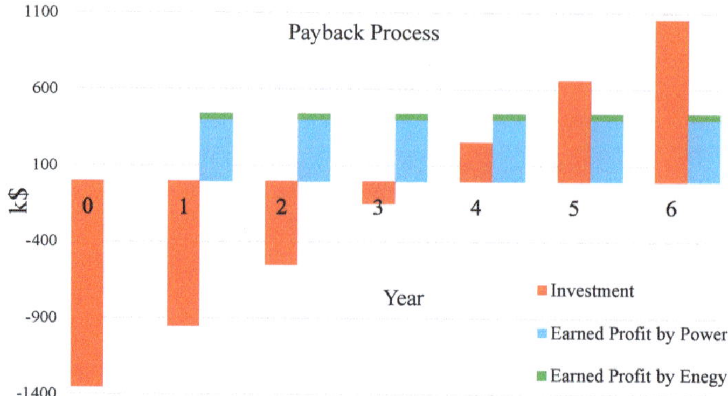

Figure 15. Payback process based on annual earning for the 5-STS crane group.

8. Conclusions

The present paper studied the techno-economic possibility of ultracapacitor (UCs) energy storage sizing in the ship to shore (STS) cranes at a container terminal (CT) port. The goal is to reduce the peak power absorbed by UC sizing. Hence, by hierarchical control, first based on a DSM strategy using the PSO algorithm along with UC sizing in the DC link for a group of STS cranes, and second by DSM strategy for each STS crane group, that goal is acheived. The proposed solution in a technical point of view has high efficiency even in temporary conditions, short term payback and suitable and efficient operation. In this regard, payback is obtained in four years. In this regard, the initial investment will be paid back before the end of the fourth year and from the fifth year onward, the investment will be profitable.

Author Contributions: Investigation, writing, review, and edit M.K.; supervision, G.P. and L.M.; project administration, B.C. All authors have read and agreed to the published version of the manuscript.

Funding: This research received no external funding.

Conflicts of Interest: The authors declare no conflict of interest.

Nomenclature

P_{ave}	Average Power
P_{max}	Maximum Power
$P^{\wedge}F$	Peak Factor
$P^{\wedge}S$	Peak Shaving
Abbreviations	
CT	Container Terminals
DC	Duty Cycle
DSM	Demand Side Management
ED	Energy Demand
ER	Energy Regeneration
FESS	Flywheel Energy Storage System
OF	Objective Function
PSO	Particle Swarm Optimization
POLB	Port of Long Beach
PTS	Platform to shore sub-crane.
STP	Ship to Platform sub-crane.
STS	Ship to Shore
TEU	Twenty-foot Equivalent Units

References

1. Parise, G.; Honorati, A.; Parise, L.; Martirano, L. Near zero energy load systems: The special case of port cranes. In Proceedings of the 2015 IEEE/IAS 51st Industrial & Commercial Power Systems Technical Conference (I&CPS), Calgary, AB, Canada, 5–8 May 2015; pp. 1–6.
2. Parise, G.; Parise, L.; Martirano, L.; Ben Chavdarian, P.; Su, C.L.; Ferrante, A. Wise port and business energy management: Port facilities, electrical power distribution. *IEEE Trans. Ind. Appl.* **2016**, *52*, 18–24. [CrossRef]
3. Homepage—Port of Long Beach. Available online: https://www.polb.com/ (accessed on 27 March 2020).
4. Port Statistics—Port of Long Beach. Available online: https://www.polb.com/business/port-statistics/#latest-statistics (accessed on 27 March 2020).
5. Wang, W.; Peng, Y.; Li, X.; Qi, Q.; Feng, P.; Zhang, Y. A two-stage framework for the optimal design of a hybrid renewable energy system for port application. *Ocean Eng.* **2019**, *191*, 106555. [CrossRef]
6. Al-Falahi, M.D.A.; Tarasiuk, T.; Jayasinghe, S.G.; Jin, Z.; Enshaei, H.; Guerrero, J.M. Ac ship microgrids: Control and power management optimization. *Energies* **2018**, *11*, 1458. [CrossRef]
7. Fang, S.; Wang, Y.; Gou, B.; Xu, Y. Toward Future Green Maritime Transportation: An Overview of Seaport Microgrids and All-Electric Ships. *IEEE Trans. Veh. Technol.* **2020**, *69*, 207–219. [CrossRef]
8. Parise, G.; Parise, L.; Malerba, A.; Pepe, F.M.; Honorati, A.; Chavdarian, P. Ben Comprehensive Peak-Shaving Solutions for Port Cranes. *IEEE Trans. Ind. Appl.* **2017**, *53*, 1799–1806. [CrossRef]
9. Kermani, M.; Parise, G.; Martirano, L.; Parise, L.; Chavdarian, B. Optimization of Peak Load Shaving in STS Group Cranes Based on PSO Algorithm. In Proceedings of the 2018 IEEE International Conference on Environment and Electrical Engineering and 2018 IEEE Industrial and Commercial Power Systems Europe (EEEIC / I&CPS Europe), Palermo, Italy, 12–15 June 2018; pp. 1–5.
10. Kermani, M.; Parise, G.; Martirano, L.; Parise, L.; Chavdarian, B. Power balancing in STS group cranes with flywheel energy storage based on DSM strategy. In Proceedings of the 2018 IEEE 59th International Scientific Conference on Power and Electrical Engineering of Riga Technical University (RTUCON), Riga, Latvia, 12–13 November 2018; pp. 1–5.
11. Kermani, M.; Parise, G.; Martirano, L.; Parise, L.; Chavdarian, B.; Su, C.L. Optimization of Energy Consumption in STS Group Cranes by Using Hybrid Energy Storage Systems Based on PSO Algorithm. In Proceedings of the 2019 IEEE Industry Applications Society Annual Meeting, Baltimore, MD, USA, 29 September–3 October 2019; pp. 1–5.
12. Chang, L.; Zhang, W.; Xu, S.; Spence, K. Review on Distributed Energy Storage Systems for Utility Applications. *CPSS Trans. POWER Electron. Appl.* **2017**, *2*, 267–276. [CrossRef]
13. Guerrero, M.A.; Romero, E.; Barrero, F.; Milanes, M.; Gonzalez, E. Supercapacitors: Alternative Energy Storage Systems. *Power Electron. Electr. Syst.* **2009**, *85*, 188–195.
14. Grbovi´c, P.J. *Ultra-Capacitors in Power Conversion Systems*; John Wiley & Sons: Hoboken, NJ, USA, 2014; ISBN 9781118356265.
15. Grbović, P.J.; Delarue, P.; Le Moigne, P.; Bartholomeus, P. The ultracapacitor-based controlled electric drives with braking and ride-through capability: Overview and analysis. *IEEE Trans. Ind. Electron.* **2011**, *58*, 925–936. [CrossRef]
16. Grbović, P.J.; Delarue, P.; Le Moigne, P.; Bartholomeus, P. A three-terminal ultracapacitor-based energy storage and PFC device for regenerative controlled electric drives. *IEEE Trans. Ind. Electron.* **2012**, *59*, 301–316. [CrossRef]
17. Bolonne, S.R.A.; Chandima, D.P. Sizing an energy system for hybrid li-ion battery-supercapacitor RTG cranes based on state machine energy controller. *IEEE Access* **2019**, *7*, 71209–71220. [CrossRef]
18. Niu, W.; Huang, X.; Yuan, F.; Schofield, N.; Xu, L.; Chu, J.; Gu, W. Sizing of Energy System of a Hybrid Lithium Battery RTG Crane. *IEEE Trans. Power Electron.* **2017**, *32*, 7837–7844. [CrossRef]
19. Zhao, N.; Schofield, N.; Niu, W. Energy storage system for a port crane hybrid power-train. *IEEE Trans. Transp. Electrif.* **2016**, *2*, 480–492. [CrossRef]
20. Iannuzzi, D.; Piegari, L.; Tricoli, P. Use of supercapacitors for energy saving in overhead travelling crane drives. In Proceedings of the 2009 International Conference on Clean Electrical Power, Capri, Italy, 9–11 June 2009; pp. 562–568.
21. Chen, D.; Niu, W.; Gu, W.; Schofield, N. Game-based energy management method for hybrid RTG cranes. *Energies* **2019**, *12*, 3589. [CrossRef]

22. Dotelli, G.; Ferrero, R.; Gallo Stampino, P.; Latorrata, S.; Toscani, S. Supercapacitor Sizing for Fast Power Dips in a Hybrid Supercapacitor-PEM Fuel Cell System. *IEEE Trans. Instrum. Meas.* **2016**, *65*, 2196–2203. [CrossRef]

23. Dimitra Tragianni, S.; Oureilidis, K.O.; Demoulias, C.S. Supercapacitor sizing based on comparative study of PV power smoothing methods. In Proceedings of the 2017 52nd International Universities Power Engineering Conference (UPEC), Heraklion, Greece, 28–31 August 2017; pp. 1–6.

24. Babazadeh, H.; Gao, W.; Lin, J.; Cheng, L. Sizing of battery and supercapacitor in a hybrid energy storage system for wind turbines. In Proceedings of the PES T&D 2012, Orlando, FL, USA, 7–10 May 2012; pp. 1–7.

25. Di Napoli, A.; Ndokaj, A. Ultracapacitor storage for a 50t capacity gantry crane. In Proceedings of the 2012 XXth International Conference on Electrical Machines, Marseille, France, 2–5 September 2012; pp. 2014–2018.

26. Ndokaj, A.; Di Napoli, A.; Pede, G.; Pasquali, M. Regulation strategy of an Ultracapacitor storage model for a gantry crane. In Proceedings of the IECON 2013—39th Annual Conference of the IEEE Industrial Electronics Society, Vienna, Austria, 10–13 November 2013; pp. 1209–1216.

27. Wu, Y.; Mi, W.J.; Tang, T.H. The design of the super capacitor group used in heavy gantry portal cranes. In Proceedings of the 2011 IEEE Power Engineering and Automation Conference, Wuhan, China, 8–9 September 2011; pp. 96–99.

28. Li, T.; Huang, L.; Liu, H. Energy management and economic analysis for a fuel cell supercapacitor excavator. *Energy* **2019**, *172*, 840–851. [CrossRef]

29. Snoussi, J.; Ben Elghali, S.; Mimouni, M.F. Sizing and Control of Onboard Multisource Power System for Electric Vehicle. In Proceedings of the 2019 19th International Conference on Sciences and Techniques of Automatic Control and Computer Engineering (STA), Sousse, Tunisia, 24–26 March 2019; pp. 347–352.

30. Sirmelis, U.; Zakis, J.; Grigans, L. Optimal supercapacitor energy storage system sizing for traction substations. In Proceedings of the 2015 IEEE 5th International Conference on Power Engineering, Energy and Electrical Drives (POWERENG), Riga, Latvia, 11–13 May 2015; pp. 592–595.

31. Hijazi, A.; Bideaux, E.; Venet, P.; Clerc, G. Electro-thermal sizing of supercapacitor stack for an electrical bus: Bond graph approach. In Proceedings of the 2015 Tenth International Conference on Ecological Vehicles and Renewable Energies (EVER), Monte Carlo, Monaco, 31 March–2 April 2015; pp. 1–8.

32. Araújo, R.E.; De Castro, R.; Pinto, C.; Melo, P.; Freitas, D. Combined sizing and energy management in EVs with batteries and supercapacitors. *IEEE Trans. Veh. Technol.* **2014**, *63*, 3062–3076. [CrossRef]

33. Kalesar, B.M.; Noshahr, J.B.; Kermani, M.; Bavandsavadkoohi, H.; Ahbab, F. Effect of Angles of Harmonic Components of Back to Back Converter of Distributed Generation Resources on Current Behavior of Distribution Networks. In Proceedings of the 2018 IEEE International Conference on Environment and Electrical Engineering and 2018 IEEE Industrial and Commercial Power Systems Europe (EEEIC / I&CPS Europe), Palermo, Italy, 12–15 June 2018; pp. 1–4.

34. Kim, S.M.; Sul, S.K. Control of rubber tyred gantry crane with energy storage based on supercapacitor bank. *IEEE Trans. Power Electron.* **2006**, *21*, 1420–1427. [CrossRef]

35. Kermani, M.; Parise, G.; Martirano, L.; Parise, L.; Chavdarian, B. Utilization of Regenerative Energy by Ultracapacitor Sizing for Peak Shaving in STS Crane. In Proceedings of the 2019 IEEE International Conference on Environment and Electrical Engineering and 2019 IEEE Industrial and Commercial Power Systems Europe (EEEIC / I&CPS Europe), Genova, Italy, 11–14 June 2019.

36. Plotnikov, I.; Braslavsky, I.Y.; Yeltsin, B.N.; Ishmatov, Z. About using the frequency-controlled electric drives with supercapacitors in the hoisting applications. In Proceedings of the 2015 International Siberian Conference on Control and Communications (SIBCON), Omsk, Russia, 21–23 May 2015; pp. 1–9.

37. Maxwell Technologies. *125 V Heavy Transportation Supercapacitor Module Datasheet*; Maxwell Technologies: San Diego, CA, USA, 2019.

38. Grbovic, P.J.; Delarue, P.; Le Moigne, P. Selection and design of ultra-capacitor modules for power conversion applications: From theory to practice. In Proceedings of the 7th International Power Electronics and Motion Control Conference, Harbin, China, 2–5 June 2012; pp. 771–777.

39. Corral-Vega, P.J.; García-Triviño, P.; Fernández-Ramírez, L.M. Design, modelling, control and techno-economic evaluation of a fuel cell/supercapacitors powered container crane. *Energy* **2019**, *186*, 115863. [CrossRef]

40. Eberhart, R.; Kennedy, J. New optimizer using particle swarm theory. In Proceedings of the Sixth International Symposium on Micro Machine and Human Science, Nagoya, Japan, 4–6 October 1995; pp. 39–43.

41. Cristian, D.; Barbulescu, C.; Kilyeni, S.; Popescu, V. Particle swarm optimization techniques. Power systems applications. In Proceedings of the 2013 6th International Conference on Human System Interactions (HSI), Nagoya, Japan, 4–6 October 1995; pp. 312–319.

MDPI

St. Alban-Anlage 66

4052 Basel

Switzerland

Tel. +41 61 683 77 34

Fax +41 61 302 89 18

www.mdpi.com

Energies Editorial Office

E-mail: energies@mdpi.com

www.mdpi.com/journal/energies

9 783039 363889